Laboratory Manual

for use with

Chemistry
A World of Choices

Second Edition

Paul B. Kelter
The University of North Carolina at Greensboro

James D. Carr
University of Nebraska

Andrew Scott

Prepared by

Jerry Walsh
The University of North Carolina at Greensboro

Dennis Burnes
The University of North Carolina at Greensboro

Gerald C. Swanson
Daytona Beach Community College
Daytona Beach, Florida

McGraw Hill

Boston Burr Ridge, IL Dubuque, IA Madison, WI New York San Francisco St. Louis
Bangkok Bogotá Caracas Lisbon London Madrid
Mexico City Milan New Delhi Seoul Singapore Sydney Taipei Toronto

McGraw-Hill Higher Education

A Division of The McGraw-Hill Companies

Laboratory Manual for use with
CHEMISTRY: A WORLD OF CHOICES, SECOND EDITION
PAUL B. KELTER, JAMES D CARR, AND ANDREW SCOTT

Published by McGraw-Hill Higher Education, an imprint of The McGraw-Hill Companies, Inc , 1221 Avenue of the Americas, New York, NY 10020 Copyright © The McGraw-Hill Companies, Inc., 2003, 1999. All rights reserved.

The contents, or parts thereof, may be reproduced in print form solely for classroom use with CHEMISTRY: A WORLD OF CHOICES, provided such reproductions bear copyright notice, but may not be reproduced in any other form or for any other purpose without the prior written consent of The McGraw-Hill Companies, Inc , including, but not limited to, in any network or other electronic storage or transmission, or broadcast for distance learning.

This book is printed on acid-free paper

   3 4 5 6 7 8 9 0 DCD DCD 0 9 8 7 6 5 4

ISBN 0-07-240162-1

www mhhe com

Acknowledgements

To Paul, for his inspiration and enthusiasm in teaching and excellence. To Dennis, for his support and cooperation on this project and many other aspects of laboratory teaching. To Jamie and Vonda, for evaluations and suggestions on experiments. To Judith, for her patient support and understanding during time spent on this project.

Jerry Walsh
Second Edition

To Paul, for his enthusiasm and excellence in teaching. To Jerry, for his support and cooperation in my quest to provide an excellent lab experience for our students. To all my students, who over the years have been a continual source of ideas. To Jo, for her continued patience during time spent on this project and her encouragement to expand my horizons.

Dennis Burnes
Second Edition

Many people deserve credit for their assistance in the production of this manual. First of all, the students of Chemistry 1025 at Daytona Beach Community College, over the period of many semesters, have provided a number of needed and appreciated comments for the improvement of the experiments. Considerable contributions have been made by innumerable lab assistants over the years in testing out many of the experiments.

To my colleagues, Fred Fathi, Jim Johnson, Ann Cooney, Frances Monroe and Al Bonamy go my indebtedness for their reviews of the experiments in the manual and their support throughout its development. A word of special thanks must go to Anne Cooney for her many excellent suggestions. Her counsels led to many improvements in the experiments and problem sets of this latest edition of the lab manual. The hand of gratitude must also be extended to DBCC for the financial assistance that led to the development of an experimental lab for Introduction to Chemistry.

My deepest appreciation I must however reserve for my spouse JoAnne, who as an educator has given me countless ideas for the improvement of this manual over the years, and labored with me in the testing of many of these experiments. Her patience must also be acknowledged for bearing with me while I sequestered myself away in front of the computer for many a late night revision of an experiment.

Gerald C. Swanson
First Edition

Preface

Chemistry is a science based on experimentation, observation, and organization of information. The true study of chemistry as a science requires the student to experience the processes by which chemical compounds are prepared, purified, evaluated, and characterized in a systematic study. The activities in this laboratory manual provide an introduction to the methods, materials, and processes used by chemists to obtain knowledge and understanding of the molecular world. Enjoy the adventure!

To voyage into the world of chemistry requires the development of a new vocabulary that deals with many things unseen and certainly many concepts never before encountered. Using this laboratory manual for **Chemistry: A World of Choices**, the student will gain first-hand experience with chemical and physical investigations. The experiments are fairly short, generally ranging from one to two hours. The normal format for the lab period will consist of a discussion session, and then the experiment.

Each experiment involves an investigation to measure or evaluate specific properties of chemical substances. Data is collected and recorded on experimental pages. Each experiment is followed by an Experimental Problems section, where questions and problems probe student understanding of and insight into the investigation. Show all your work when answering mathematical questions. Please write neatly and legibly. These pages can be torn out, stapled, and turned in to the instructor.

The performance of any experiment is always rooted in a clear understanding of the theory and the technique. It is therefore important that the student prepare for the experiment before entering the lab. Much less time will then be spent by the student trying to puzzle through the experiment at the bench top, or making needless errors based on a lack of a clear understanding. Each experiment has been tested many times for ease of operation and, with proper preparation, the student should find them both interesting and informative.

Contents

	Page
Acknowledgments	iii
Preface	iv
Safety in the Laboratory	vii
Laboratory Equipment and Techniques	ix

Experiments

1. Physical and Chemical Changes	1
2. Elements, Compounds, and Mixtures	7
3. The Graphing of Data	19
4. Measurements of Length and Mass	29
5. The Measurement of Density	35
6. Periodic Properties of Elements	43
7. Solubility and Miscibility	55
8. The Syntheses and Properties of Carbon Dioxide, Oxygen, and Hydrogen	61
9. The Synthesis and Properties of Ammonia	69
10. Precipitation Reactions and Filtration	75
11. The Qualitative Analysis of an Ionic Compound	87
12. The Determination of the Molar Weight of Butane	95
13. Chemical Equilibrium and Le Chatelier's Principle	103
14. The pH of Common Substances	113
15. Buffers and pH Changes	119
16. Electrolysis and Electroplating	125
17. The Synthesis and Recrystallization of Aspirin	135
18. The Extraction of Caffeine from Coffee	143
19. The Preparation of Soap	147
20. Radioactivity	153
21. Analysis of Acids	163
22. Characteristics of Antacids	169

Appendix A. Periodic Table of the Elements	176
Appendix B. Conversion Factors	177

Safety

The following is a guide to provide you with a safe and enjoyable lab experience.

1. Eye protection (safety glasses, goggles, or face shields) must be worn by all students when working in the laboratory.

2. No food or drink is allowed in the laboratory.

3. Do not taste any chemical.

4. Clothing, purses, books, briefcases, or backpacks should not be placed on your bench top or on the floor next to your work area. In an emergency, these may provide an obstacle course to a safe exit. Instead, place these items out of the way as directed by your instructor.

5. Bare feet, sandals, and open toed shoes are prohibited in the laboratory. To avoid chemical burns, wear old, non-synthetic clothing. For maximum protection, wear a lab coat or apron while in the lab.

6. Long hair should be tied back or pinned up, so it will not fall into chemicals or flames.

7. If an accident occurs in the laboratory, report it to the instructor immediately.

8. If you should have skin contact with any chemical, wash it off with water at the sink immediately.

9. If you spill a large amount of corrosive chemical on your skin, you may need to use the lab safety shower. Make sure you know its location in the lab. If the chemical has gotten into your eyes, the eye-wash fountain may be used to rinse out the chemical. Know the location of the eye-wash fountain.

10. You should be aware of the location of the fire extinguishers. In case of fire, direct the extinguisher's discharge at the base of the flames. If a large fire occurs, call the fire department immediately and alert others to evacuate.

11. If you spill any chemical, solid or liquid, be sure to clean it up immediately so that another student does not come into contact with it, and perhaps be injured.

12. Before leaving the laboratory, make sure your equipment and chemicals are put away. Wipe off the desk top with a damp paper towel. Finally, wash your hands with soap and water.

13. Never perform an unauthorized experiment.

14. When observing a chemical reaction in a test tube, beaker, or flask, always observe from the side of the container. You never know if the contents may suddenly be ejected like a geyser. This is especially true if you are heating the reaction mixture.

15. Never point the open end of a test tube at yourself or another person.

16. If you have heated any glassware, be cautious when it is handled. Hot and cold glass looks exactly the same.

17. When inserting glass tubing into a stopper, lubricate the tubing with water or glycerol, wrap the tubing with a towel, and hold the end of the tubing about 1 cm from its end. Insert the tubing with a twisting motion while continuing to keep your hand close to the stopper.

18. Take only small, necessary quantities of **stock chemicals** (large bottles of chemicals that are used by the entire class). Never return any chemical back to its stock bottle from one of your containers.

19. If you want to smell a substance, do not hold it directly to your nose; instead, hold the container a few centimeters away and use your hand to fan the vapors to you. This procedure is referred to as **wafting**.

20. Avoid handling more than one **reagent** (another name for a chemical used in an experiment) bottle at a time, so that you do not interchange their stoppers by mistake. Make sure you read the labels on reagent bottles carefully. Mixing the wrong chemicals may be dangerous.

21. Always pour concentrated acids and bases **into** water when diluting them. The reverse procedure can lead to a violent boiling and splattering of the acid or base solution.

22. When disposing of liquid chemicals in the sink, flush them with large quantities of water.

23. Do not dispose of matches, paper or solid chemicals in the troughs or sink. Discard them instead into a waste bucket.

24. Dispose of any broken glassware in the container designated for broken glass.

25. Always come to lab prepared. Read the experiment before you come to lab so that you have a general idea of what you will be during the lab period.

26. Follow directions for safely disposing of chemicals.

Laboratory Equipment and Techniques

I. OBJECTIVES:

    A. To recognize the components of the Bunsen burner and to learn how they can be adjusted to control the type of flame produced.

    B. To become familiar with common laboratory equipment.

    C. To learn how to make measurements of mass and volume.

II. DISCUSSION:

    A. The Bunsen Burner:

The most common heating device in the laboratory is the Bunsen burner. A typical Bunsen burner is pictured below.

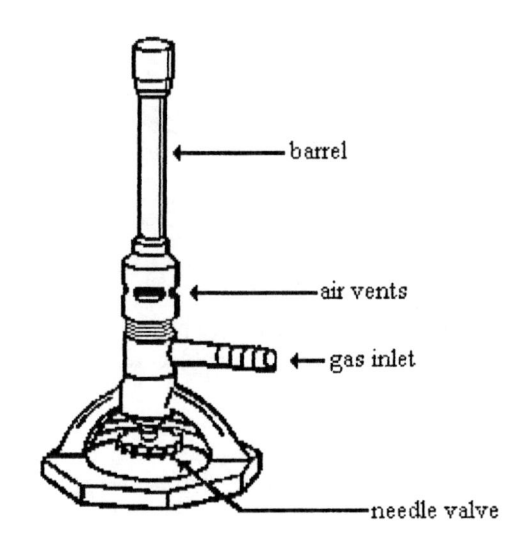

The function of the four components listed above are as follows:

1. **Needle valve:** by turning this you can control the amount of methane gas going into the burner.

2. **Gas inlet:** gas enters the burner via this inlet.

3. **Air vents:** air enters the burner via these vents.

4. **Barrel:** by rotating this you can control the amount of air going into the burner. Turning the barrel counterclockwise increases the amount of air.

The Bunsen burner used in most labs employs natural gas (primarily methane) as its fuel. The amount of fuel and air that are combined can affect the type of flame that is produced. A properly adjusted flame should consist of a total flame height of about 8 - 10 cm and two cones of combustion- a larger, light blue cone and a smaller, darker blue cone. The hottest point of the flame is at the top of this smaller, inner cone. When glassware needs to be heated very vigorously, it should be held at the top of this inner cone of flame. If the flame is too large or small, it may be adjusted by turning the needle valve.

A poorly adjusted flame sometimes has too much fuel and not enough air. This is termed a **fuel-rich flame** and is characterized by being yellow or orange in color and of relatively low temperature. By turning the barrel of the burner, more air can be admitted into the gas mixture and a blue flame obtained.

B. Heating Samples of Chemicals:

Frequently we will want to heat a chemical in a test tube to cause it to undergo a chemical change. The test tube, never more than one-third full, should **not** be held in one's hand, but rather a test tube clamp or a ring stand clamp. Put the clamp near the top of the test tube, incline the tube at about a 45° angle, and heat near the bottom of the tube cool flame. Remember to make sure that the tube is not pointed at yourself or anyone near you.

C. Measuring Mass:

The Electronic Digital Balance:

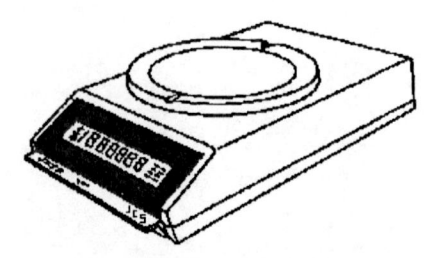

The bar on the front of this balance turns it on when pressed downward. If the balance is already on and does not read 0.00, press the bar to rezero it. The object to be weighed is placed on the pan and its mass is displayed on the digital readout.

Since chemicals cannot be placed directly on the pan of a balance, they must be weighed out in a container of some sort, such as a watch glass, beaker, or flask, or a piece of weighing paper. The chemical therefore, must be **weighed by difference**. The container is weighed first and this value is recorded **(the tare weight)**. The chemical is placed in a container and the new weight is determined. The difference in the two weights is the weight of the chemical.

A simpler procedure can be employed with the digital balance. After placing the empty container on the pan of the balance, press the bar to set the tare weight to zero. As the chemical is now added to the container, the weight of the chemical is digitally displayed.

D. Measuring volume:

1. Approximate volumes of liquids can be measured using **beakers** and **Erlenmeyer flasks** of the appropriate size. Generally, the accuracy of the measurement is ± 10 %.

2. For more accurate measurements, **graduated cylinders** are used. When making a reading the liquid level in a graduated cylinder it is important to position the cylinder properly. The reason for this is that the

surface of the liquid has a concave curvature due to the tendency of the liquid to cling to the sides of the cylinder. This produces what appears to be a layer on the top of the liquid. This "layer" is termed the **meniscus**. For ease and consistency, the bottom of the meniscus is always read. If the meniscus is viewed from different positions, different volume values can be read. This is termed **parallax**. Again, in order to get a consistency in volume readings, parallax is avoided by always reading the volume with one's eye level with the bottom of the meniscus.

E. Handling Reagents and Glassware:

1. Stock Reagent Bottles:

Most of the chemicals (frequently referred to as **reagents**) in the laboratory will be used by the entire class of students, i e., they are **stock reagents**. It is very important therefore, that they not get contaminated. If you have measured out too much of a stock reagent, discard the excess properly. **Do not put the excess reagent back into the reagent bottle**.

2. Transferring Chemicals:

a. Solids:

Small amounts of solids can sometimes be conveniently poured out of a bottle. Remove the cap or stopper of the bottle and place it upside down on the lab bench. Control of the pouring process is accomplished by tilting the bottle and rotating it at the same time. In this way very small amounts of a solid can be accurately dispensed. If the substance is somewhat sticky and cannot be poured, it can be scooped out by the use of a spatula, scoopula, or flat wooden stick that is provided with the reagent. Make sure that you do not exchange the spatula of one bottle with the spatula of another bottle. Frequently you will need about 1 or 2 grams of a solid for an experiment. It will not be necessary to weigh this out. You can estimate a gram of a solid to generally possess the volume of a small pea for most of the substances with which you will be working.

b. Liquids:

If a large amount of liquid is required, you may need to measure it out in a graduated cylinder. Remove the cap and set it upside down, on the lab bench top. Hold a glass stirring rod against the mouth of the bottle while tipping it, and allow the liquid to run down the glass rod into the graduated cylinder.

Many times you will need small, approximate amounts of liquids. Again, it will be unnecessary to measure out these volumes. Many of the reagent bottles will have an eye dropper (called a **pipette** in science labs) attached to them. A convenient rule of thumb for volume dispensing is: **one dropperful is approximately 1 mL**.

F. Disposal of Waste Chemicals:

Most of the chemicals you will work with in an introductory chemistry lab are safe enough to dispose of down the drain or in the trash container. A few substances cannot be disposed of in this manner. The instructor will provide special containers for the disposal of these substances.

Name:_____   Section:_____

EXPERIMENT 1

Physical and Chemical Changes

I. OBJECTIVES:

A. To mix chemicals and observe the resulting changes.

B. To classify these changes as chemical or physical.

II. DISCUSSION:

Properties of substances can be characterized as either **PHYSICAL** or **CHEMICAL**.

A **physical property** is one that can be observed without changing the chemical composition of a substance. Physical properties describe appearance and measurable properties that do not involve producing another substance. For example, typical physical properties are: color, odor, melting point, boiling point, mass, density, ability to conduct heat, etc. Silver, for instance, is observed to be a bright, shiny metal that is quite dense, and can conduct heat quite easily. Glass is observed to be brittle and can easily shatter.

Ice is a colorless solid with a density of 0.917 g/mL at $0°C$, and turns to a colorless liquid with a density of 0.99987 g/mL at $0°C$. Ice melting is a **physical change** because only the physical appearance of the water has changed, not the arrangement of hydrogen and oxygen atoms in the water molecule. In noting that water changes from a colorless solid to a colorless liquid when it melts, you are observing a physical change.

Any substance may also be described by its **chemical properties**, i.e., those properties which indicate the possible chemical changes a substances may undergo. For instance, a piece of coal has the chemical property of being able to burn. The coal reacts with oxygen to produce water and carbon dioxide. The arrangement of carbon atoms in the coal has been altered as they combined with oxygen atoms, and so in this example, a **chemical change** has occurred.

A chemical change is usually accompanied by the release or absorption of energy, and is usually not easily reversed. When gasoline burns, it produces water, carbon dioxide, and energy. These products are quite different from the original gasoline.

Since chemical changes produce substances with physical properties different from those of the starting materials, we can usually tell if a change is chemical by noting how physical appearances change. In the following experiment, you will use the criteria chemists very often use to make at least a preliminary judgement about the nature of some observed change. These include:

1. Is heat evolved when two substances are mixed?
2. Are bubbles formed, indicating that a gas is being produced?
3. Is there a permanent change in color?
4. Is there a permanent change in odor?
5. Does a solid substance (a precipitate) form when two solutions are mixed?

Experiment 1 Physical and Chemical Properties

In general, if a change that you have observed is reversible, that change is almost always a physical one. If you can answer "yes" to any of the above five questions, you are observing a chemical change.

III. MATERIALS:

A test tube rack, 8 small dry test tubes, 2 rubber stoppers, , a squeeze bottle of distilled water, spatulas, and miscellaneous stock chemicals, aluminum, copper chloride solution, copper wire, silver nitrate solution, 3 M hydrochloric acid, zinc, hexane, xylene, sodium hydroxide solution, sodium chloride solution

IV. PROCEDURE:

A. Mixing Aluminum with Copper Chloride:

To a small test tube add about 2 mL of a copper chloride (**CuCl₂**) stock solution. To this solution add a small piece of aluminum (**Al**). Allow this mixture to stand undisturbed for a few minutes while you perform the next few experiments. Periodically examine the mixture. Record your observations in **Table 1**, and based on the criteria in the Discussion section, indicate whether a physical or chemical change occurred. Discard the mixture in the container provided by the instructor and wash the test tube with soap and water and rinse with distilled water.

B. Mixing Copper with Silver Nitrate:

To a clean, **dry** test tube add about 2 mL of silver nitrate (**AgNO₃**) stock solution. To this solution add a small piece of copper (**Cu**) wire. Allow this mixture to stand undisturbed for a few minutes while you perform the next few experiments. Periodically examine the mixture. Record your observations in **Table 1**, and based on the criteria in the Discussion section, indicate whether a physical or chemical change occurred. Discard the mixture in the container provided by the instructor and wash the test tube with soap and water and rinse with distilled water.

C. Mixing Copper with Hydrochloric Acid:

To a clean test tube add about 2 mL of hydrochloric acid (**HCl**). To this solution add a small piece of copper (**Cu**) wire. Record your observations in **Table 1**, and based on the criteria in the Discussion section, indicate whether a physical or chemical change occurred. Discard the mixture in the container provided by the instructor and wash the test tube with soap and water and rinse with distilled water.

D. Mixing Zinc and Hydrochloric Acid:

To a small, **dry** test tube add a pea-sized amount of zinc (**Zn**). To this add about 1 mL of hydrochloric acid (**HCl**). Record your observations in **Table 1**, and based on the criteria in the Discussion section, indicate whether a physical or chemical change occurred. Discard the mixture in the container provided by the instructor and wash the test tube with soap and water and rinse with distilled water.

E. Mixing Hexane and Xylene:

To a small, **dry** test tube add about 1 mL of hexane (**C₆H₁₄**) stock reagent and about 1 mL of xylene (**C₈H₁₀**) stock reagent. Stopper the test tube with a rubber stopper and shake gently a few times. Record your observations in **Table 1**, and based on the criteria in the Discussion section, indicate whether a physical or chemical change occurred. Discard the mixture in the container provided by the instructor and wash the test tube with soap and water and rinse with distilled water.

Experiment 1 Physical and Chemical Properties

F. Mixing Hexane and Water:

To a small test tube add about 1 mL of water (**H₂O**) and about 1 mL of hexane (**C₆H₁₄**) stock reagent. Stopper the test tube and shake gently a few times. Record your observations in **Table 1**, and based on the criteria in the Discussion section, indicate whether a physical or chemical change occurred. Discard the mixture in the container provided by the instructor and wash the test tube with soap and water and rinse with distilled water.

G. Mixing Hydrochloric Acid and Sodium Hydroxide:

To a small test tube add about 1 mL of hydrochloric acid (**HCl**). To this add about 1 mL of a sodium hydroxide (**NaOH**) stock solution. Do you **see** any change occurring? Is there any other evidence of a change? Record your observations in **Table 1**, and based on the criteria in the Discussion section, indicate whether a physical or chemical change occurred. Discard the mixture in the sink and wash the test tube with soap and water and rinse with distilled water.

H. Mixing Sodium Chloride and Silver Nitrate:

To a small test tube add about 1 mL of sodium chloride (**NaCl**) stock solution. To this add about 1 mL of silver nitrate (**AgNO₃**) stock solution. Record your observations in **Table 1**, and based on the criteria in the Discussion section, indicate whether a physical or chemical change occurred. Discard the mixture in the container provided by the instructor and wash the test tube with soap and water and rinse with distilled water. Return the equipment to the designated area.

Table 1

SUBSTANCES MIXED	OBSERVATION	PHYSICAL OR CHEMICAL CHANGE
Aluminum with Copper Chloride		
Copper with Silver Nitrate		
Copper with Hydrochloric Acid		
Zinc with Hydrochloric Acid		
Hexane with Xylene		
Hexane with Water		
Hydrochloric Acid with Sodium Hydroxide		
Sodium Chloride with Silver Nitrate		

Experiment 1 Physical and Chemical Properties

Experiment 1 Physical and Chemical Properties

Name:_____   Section:_____

Experiment 1 Problems

1. What is a reagent?

2. What is a stock reagent?

3. If you need to measure out about 1 mL of a liquid, what would be a convenient, rough estimate of this volume?

4. If you need to measure out about 1 g of a solid, what would be a convenient, rough estimate of this mass?

5. A test tube should never be pointed toward yourself or someone else while it is being heated. Why?

6. What are two chemical properties of hydrochloric acid?

7. What are two chemical properties of silver nitrate?

Experiment 1 Physical and Chemical Properties

8. When a solution of sodium carbonate is added to a solution of calcium chloride, a white solid is formed. What kind of change is this? What is your reason for this choice?

9. If sugar is dissolved in water, does a chemical or physical change occur? What evidence could you present for your answer?

10. Are the following changes chemical or physical?

a. lead melting _____

b. a wooden log burning _____

c a glass bottle breaking _____

d. food spoiling _____

e. dew evaporating _____

f. starch being changed to sugar by the enzymes in saliva _____

g. a lead weight sinking to the bottom of a bottle of water _____

h. a gas being produced when vinegar and baking soda are mixed _____

i. milk souring _____

j. aluminum corroding _____

k. a spring compressing _____

l. gasoline being distilled from crude petroleum _____

m. sodium azide rapidly converting to sodium metal and nitrogen gas _____

Name:_____   Section:_____

EXPERIMENT 2
Elements, Compounds, and Mixtures

I. OBJECTIVES:

A. To observe some properties of mixtures, elements, and compounds.

B. To separate the components of mixtures by use of physical methods.

C. To combine elements to produce compounds, and then observe their properties.

D. To decompose compounds into elements, and then observe their properties.

E. To react an element with a compound, and observe a property of the product.

II. DISCUSSION:

The substances that we see around us fall into three general categories:
1. **Elements**: the fundamental type of matter that cannot be decomposed into any simpler substance.
2. **Compounds**: substances that are composed of elements, and can be decomposed into elements by chemical methods (but not by physical methods).
3. **Mixtures**: combinations of elements and/or compounds that can be separated into their components by physical methods.

In this experiment, you will examine these three types of matter.

Physical methods that are commonly employed to separate the components of mixtures include filtration, evaporation, distillation, condensation, crystallization, magnetic attraction, and centrifugation. All these techniques involve separating existing substances, while not creating anything new that was not present already. You will separate a salt-water mixture and a sand/iron mixture by using some of these techniques.

You will also examine the combination of elements to form compounds. This is a chemical change that produces a new substance with physical properties different from the original reactants. By reacting iron with sulfur, you will produce the compound iron(II) sulfide and determine some of its properties.

$$Fe + S \rightarrow FeS$$

Likewise, upon heating you will be able to unite copper and oxygen to produce a compound, copper(II) oxide, which has drastically different properties from the starting copper and oxygen.

$$2\,Cu + O_2 \rightarrow 2\,CuO$$

Experiment 2 Elements, Compounds, and Mixtures

In a similar fashion, you will be able to unite magnesium with oxygen to give the compound, magnesium oxide.

$$2\,Mg + O_2 \rightarrow 2\,MgO$$

The properties of these substances will also be examined.

Compounds can be decomposed to elements or other compounds. You will examine the decomposition of hydrogen peroxide into water and oxygen.

$$2\,H_2O_2 \rightarrow 2\,H_2O + O_2$$

A simple test will be conducted to determine the presence of oxygen.

Elements can also react with compounds. By adding calcium metal to water, you will observe a physical change and the production of a new substance, hydrogen gas.

$$Ca + 2\,H_2O \rightarrow Ca(OH)_2 + H_2$$

The formation of $Ca(OH)_2$, calcium hydroxide, can be detected by the use of litmus paper. Calcium hydroxide is a member of a group of compounds called **bases** (compounds that produce hydroxide ions, OH^-, in solution). Their presence can be detected by the use of an **indicator**, a substance that changes color depending on the acidic or basic nature of a solution. The indicator you will use is called litmus. It is usually found in the lab as small strips of paper impregnated with the indicator. A base can be detected by its ability to change red litmus paper to a blue color. You will use this to determine the presence of the hydroxide ion

III. MATERIALS:

50 mL beakers, 10 mL graduated cylinder, electronic balance, wire gauze, iron ring, Bunsen burner, striker, two 4" test tubes, magnet, test tube containing a sand/iron unknown, porcelain crucible with lid, spatula, clay triangle, steel wool, watch glass, forceps, 6" test tube, buret clamp, ring stand, wooden splint, sand paper, red and blue litmus paper, sodium chloride, silver nitrate solution, iron filings, powdered sulfur, 6 M hydrochloric acid, copper wire, magnesium ribbon, 3% hydrogen peroxide, potassium iodide, calcium turnings.

IV. PROCEDURE:

A. The Properties and Separation of a Salt Water Solution:

1. Tare a 50 mL beaker and record its mass in **Table 1**. To the beaker add about two grams of sodium chloride **(NaCl)**. Weigh the beaker again so that you get an accurate value for the total mass and record this in the Table. By difference, determine the mass of sodium chloride and record this in the Table.

2. To the beaker, add 10 mL of distilled water. Swirl the mixture until all the salt has dissolved.

3. Place the beaker on a wire gauze supported on an iron ring, and heat the salt solution with a Bunsen burner. Continue heating until all the water boils away, leaving only a white solid. When no liquid remains, discontinue heating and allow the beaker to cool.

4. After the beaker has cooled to room temperature, reweigh it and record its mass in the Table. By difference, determine the mass of the solid that remains in the beaker. Record this in the Table. How does this compare to the original mass of the sodium chloride?

Experiment 2 Elements, Compounds, and Mixtures

Answer: _____

5. In order to test this solid and see if it is salt, place a few crystals in a 4" test tube. To a second test tube, add a few crystals of sodium chloride from the reagent bottle. Add about a milliliter of distilled water to each test tube, and agitate the tubes until the crystals have dissolved. To each test tube, add a drop of silver nitrate **(AgNO₃)** solution. In the Table, record your observations of what occurs in each test tube.

6. Dispose of the contents of the test tubes in the container designated by the instructor. Wash out the test tubes and beaker with soap and water and rinse them with distilled water.

Table 1

	Data/Observations
Tare Weight of the 50 mL Beaker	
Mass of the Beaker and NaCl	
Mass of the NaCl	
Mass of the Beaker and White Solid	
Mass of the White Solid	
Solution of White Solid + AgNO₃	
Solution of NaCl + AgNO₃	

B. The Properties and Separation of a Sand/Iron Mixture:

1. Obtain a magnet and an unknown consisting of a mixture of iron **(Fe)** and sand **(SiO₂)**. Record the number of the unknown in **Table 2**.

2. Tare a dry beaker **(Beaker 1)** and record its weight in the Table. Transfer the iron/sand mixture to the beaker. Reweigh the beaker and record its mass in the Table. By difference, calculate the mass of the mixture of iron and sand, and record this value in the Table.

3. Bring the magnet next to the side of the beaker. What properties of iron and sand do you observe?

Answer: _____

4. Tare a second dry beaker **(Beaker 2)** and record its mass. Use the property you observed above to separate the iron from the sand. Transfer the iron to the tared beaker and reweigh the beaker. Record this value in the Table. By difference, determine the mass of the iron and record this value in the Table.

Experiment 2 Elements, Compounds, and Mixtures

5. Calculate the weight percentage of iron in the unknown that you obtained.

$$\text{Weight Percentage of Iron} = \frac{\text{Grams of Iron}}{\text{Grams of Mixture}} \times 100\%$$

Record this value in the table.

6. Pour the iron filings and sand back into the test tube with the sand and return the unknown to the instructor.

Table 2

	Data
Unknown Number	
Tare Weight of Beaker 1	
Mass of Beaker and Mixture	
Mass of the Mixture	
Tare Weight of Beaker 2	
Mass of Beaker and Iron	
Mass of the Iron	
Weight Percentage of the Iron in the Mixture	

C. The Combination of Iron and Sulfur:

1. Obtain a porcelain crucible and lid. To the crucible, add 2 g of iron **(Fe)** filings and 2 g of powered sulfur **(S)**. With a spatula, stir the mixture thoroughly to mix the contents.

2. Place the crucible on a clay triangle on an iron ring in a hood and cover it with the lid. With a Bunsen burner, heat the mixture red hot for about three minutes.

3. Allow the crucible to cool to room temperature. Examine the product in the crucible and describe its appearance in **Table 3**.

4. Break off a small piece of this substance and test it with a magnet. Record your observation in the Table. Place this piece of product in a 4" test tube. To a second test tube, add a small amount of iron filings.

5. Carry out the next part of the procedure in the hood. To each test tube, add a milliliter of 6M hydrochloric acid **(HCl)** and record your observations in the Table.
Caution: Hydrochloric acid is corrosive. If you get any on your skin, wash it off immediately with copious amounts of water. Record you observations.

6. Dispose of the contents of the test tubes and the crucible in the container designated by the instructor and wash the equipment with soap and water.

Table 3

	Observations
Reaction Product of Iron and Sulfur	
Test of Product with Magnet	
Product Tested with Hydrochloric Acid	
Iron Tested with Hydrochloric Acid	

D. The Combination of Copper and Oxygen:

1. Clean a piece of copper **(Cu)** wire with steel wool and observe its appearance. Record your observations in **Table 4**.

2. Heat the copper wire red hot in the flame of a Bunsen burner. A coating will appear on its surface. Remove the wire from the flame and carefully scrape off the coating with a spatula onto a watch glass.

3. Repeat the heating and scraping several more times until you have a small pile of the coating on the watch glass. Observe its appearance, and record your observation in the Table.

Table 4

	Observations
Appearance of the Clean Copper	
Appearance of the Coating	

Experiment 2 Elements, Compounds, and Mixtures

E. The Combination of Magnesium and Oxygen:

1. Clean a small (~5 cm) piece of magnesium (**Mg**) ribbon with steel wool. Observe its properties and record your observations in **Table 5**.

2. Hold the end of the ribbon with a pair of forceps and heat the other end in the flame of a Bunsen burner. The ribbon will begin to burn.

Caution: Do not look directly at the burning ribbon. The emitted light is extremely bright and can cause eye damage.

3. As soon as the ribbon begins to burn, position it directly over a watch glass. When it stops burning, place the residue on the watch glass and observe its properties. Touch it with the tip of a spatula. Record your observations in the Table.

Table 5

	Observations
Clean Magnesium Ribbon	
Magnesium Combustion Product	

F. The Decomposition of Hydrogen Peroxide

1. In a 6" test tube, add 5 mL of 3% hydrogen peroxide (H_2O_2). Clamp the test tube to a ring stand.

2. Add about 1 g of potassium iodide (**KI**) to the test tube. Record your observations in **Table 6**.

3. Light a wooden splint and allow it to burn vigorously. Blow out the flame so that the splint is glowing. Plunge the splint into the opening of the test tube. Record your observations in the table.

4. Discard the hydrogen peroxide solution. Wet the splint with water and discard it. Wash and rinse the test tube.

Experiment 2 Elements, Compounds, and Mixtures

Table 6

	Observations
Hydrogen Peroxide and Potassium Iodide	
Glowing Splint in Test Tube	

G. The Reaction of Calcium with Water:

1. Clean a small piece of calcium (**Ca**) with some sandpaper. Record your observations of the calcium's properties in **Table 7**.

2. Add the calcium turning to a 50 mL beaker containing about 25 mL of water. Allow the reaction to occur for a few minutes. What do you see occurring? Record your observation in the Table.

3. By using a piece of red and blue litmus paper, test the acidity/basicity of distilled water. Using fresh pieces of red and blue litmus paper, test the water in your beaker containing the calcium. Record your observations in the Table. Dispose of the calcium in the container provided by your instructor. Wash your glassware with soap and water, and then return them to the designated area.

Table 7

	Observations
Freshly Cleaned Calcium	
Calcium Added to the Water	
Blue Litmus in Distilled Water	
Red Litmus in Distilled Water	
Blue Litmus in the Calcium/Water Mixture	
Red Litmus in the Calcium/Water Mixture	

Experiment 2 Elements, Compounds, and Mixtures

Experiment 2 Elements, Compounds, and Mixtures

Name:_____   Section:_____

Experiment 2 Problems

1. What is a mixture? How does it differ from elements and compounds?

2. List common physical methods of separating the components of mixtures.

   a.   d.

   b.   e.

   c.   f.

3. What <u>physical method</u> could be used to separate the components of the following mixtures?

   a. blood (remove the red blood cells)

   b. carbonated beverage (remove the CO_2)

   d. sugar water

   e. sand and water

4. How does a compound differ from an element?

5. Write an equation that describes the following chemical reactions.

   a. When hydrogen peroxide (H_2O_2) decomposes, it produces water and oxygen.

Experiment 2 Elements, Compounds, and Mixtures

    b. Nitrogen and hydrogen unite to give ammonia (NH_3).

    c. When electricity is passed through water, it decomposes to give oxygen and hydrogen.

    d. Sulfur burns in the presence of air to produce sulfur dioxide (SO_2).

6. Some chemical reactions give off energy when they occur. These are termed **exothermic**. Other chemical reactions absorb energy when they occur. These are termed **endothermic**. Identify the exothermic and endothermic reactions in this experiment.

          Exothermic   **Endothermic**

7. Identify the following as exothermic or endothermic reactions.

    a. gasoline burning in an auto engine _____

    b. photosynthesis _____

    c. water being electrolyzed to hydrogen and oxygen _____

    d. dynamite exploding _____

    e. when sulfuric acid is mixed with sodium hydroxide, the resulting solution gets hot _____

8. A substance is believed to be an acid. How could you easily test it to prove it was acidic? What would you see when you performed the test?

Experiment 2 Elements, Compounds, and Mixtures

9.a. What gas was evolved from the calcium metal when it was placed in water? _____

   b. How could you test this gas to determine its identity?

Experiment 2 Elements, Compounds, and Mixtures

Name:_____   Section:_____

EXPERIMENT 3

The Graphing of Data

I. OBJECTIVES:

A. To learn the terms associated with a graph.

B. To learn how to properly graph data.

C. To construct different types of graphs, given numerical data.

II. DISCUSSION:

When large amounts of data are collected in the laboratory, the relationship of one set of data to another may not be very clear. For example, as a gas is heated, how is the pressure affected? To more clearly represent the relationship of these data, a graph is frequently constructed, helping to see a trend that was not initially obvious.

Normally, when measurements are made, some property is controlled and changed by the experimenter; for example, the temperature of the gas. This is, in general, termed the **independent variable**. In response to a change made in this property by an experimenter, some other property changes and is measured; the pressure of the gas, for example. This is termed the **dependent variable**.

After obtaining the data, a graph can be constructed. It consists of two perpendicular axes. The horizontal axis (X-axis) is termed the abscissa, while the vertical axis (Y-axis) is the ordinate. In the above example of pressure and temperature, temperature would be plotted on the X-axis, and pressure on the Y-axis. In making a graph, some general considerations should be kept in mind.

A. Choosing the Axes:
Normally, the dependent variable is plotted on the Y-axis, with the independent variable on the X-axis. Graph paper is frequently 8 1/2" x 11", which makes the graph itself rectangular. Sometimes the shorter side is used for the X-axis, and at other times it is used for the Y-axis. The orientation of the graph is chosen so that the data is spread over much of the graph and axes have convenient scales

B. Labeling the Axes:
Label both the X- and Y-axis with the appropriate names of the variables, followed by a comma, and then the units of the variables.

C. Choosing Appropriate Scales for the Axes:
A set of data should span most of the axis on which it is plotted. If the range of temperatures measured went from 22°C to 95°C, then the X-axis should have a scale that reads from 20°C to 100°C. This choice of range however, is also dependent on the divisions on the scale. In order to make the graphing of the data easy, a scale should be chosen in which each division corresponds to a value of 0.5, 1, 2, 5, 10, etc.

Experiment 3 The Graphing of Data

An odd value for a division, such as, 3.33 would make graphing tedious and lead to errors. Some examples of scales that are easy to use are shown in **Figure 1**.

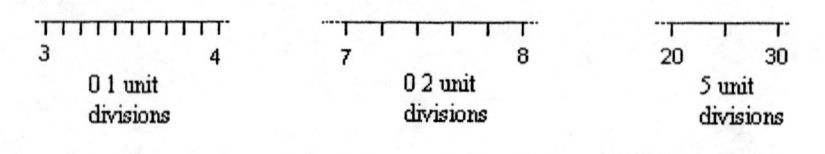

```
 |||||||||||         |  |  |  |  |  |        |    |    |
 3         4         7              8        20        30
   0.1 unit              0.2 unit              5 unit
   divisions             divisions             divisions
```

Figure 1

Rather awkward scales that would be difficult to use are shown in **Figure 2**.

```
  |  |  |  |          |||||||
  9       10         80       90
   0.33 unit            1.7 unit
   divisions            divisions
```

Figure 2

The choice of the divisions on one scale need not be the same as those on the other scale.

The choice of scale divisions is also important if you wish to interpolate a point on the graph, i.e., determine the X- or Y-value of a point on the curve that has been drawn. To do this, a vertical or horizontal line would be drawn from a point on the curve to the appropriate axis. At the intersection of the line on the axis, the value may be read. If an awkward scale was chosen, it would be difficult to read the value of this intersection.

D. Plotting the Data:
To plot pairs of data as points on the graph, find the value of an independent variable (on the X-axis) and then move vertically upward until the value of the dependent variable is reached (on the Y-axis). At that point, place a pencil mark, and draw a small circle around it.

E. Drawing a Curve:
After plotting all the points from the data, a curve should be drawn that represents the best fit for the data. This means that all the points may not be on the line, but as many as possible are as close as possible to the curve.

If the points appear to lie roughly in a straight line, a ruler may be used to draw the line. Lines should not be drawn from point to point, but rather a continuous straight line should be drawn that best fits the entire set of points. An example is shown in **Figure 3**.

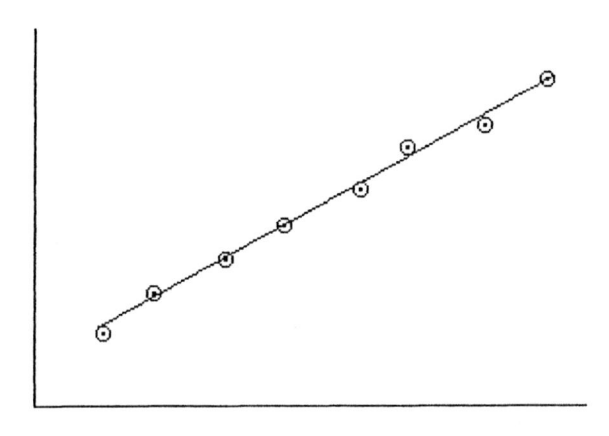

Figure 3

If the points do not appear to lie in a straight line, a smooth curve should be drawn that best fits these points. If available, a French curve may be used to draw the curve. This device is piece of plastic with a multitude of curves that can be turned in different ways until one of the curves matches part of the points on the graph. Again, not all the points may lie on the curve. **Figure 4** illustrates this type of curve.

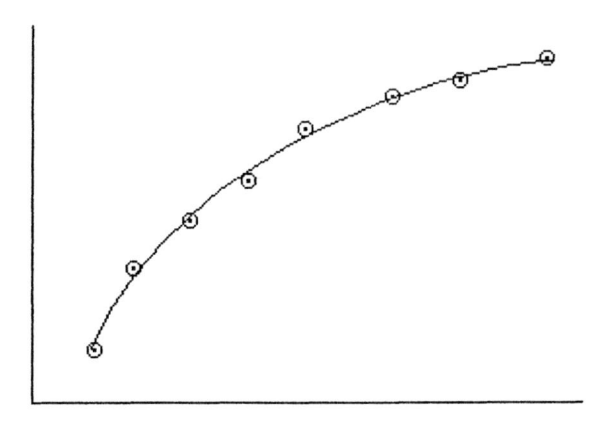

Figure 4

F. Titling the Graph:
On the upper part of the graph, away from the curve that has been drawn, place a title that describes what has been plotted. For example: "The Relationship of the Pressure of a Gas to Its Temperature." Also, place your name at the very top of the page, where it can easily be seen.

G. Interpolating Data:
Often the prediction of an X value from a measured Y value is desired, or vice versa. To interpolate X or Y values, draw a horizontal or vertical line from the axis corresponding to the known value to the curve. Then draw a vertical or horizontal line from that point on the curve to the other axis. Read the intersection of the line with the axis as accurately as possible.

Experiment 3 The Graphing of Data

H. Determining the Slope of a Straight Line:
If a straight line can be drawn from the data, the change of the dependent variable, relative to the independent variable, can be determined from the slope of the line. For example, if the pressures of a gas are plotted, versus their temperatures, a straight line is obtained, as shown in **Figure 5**.

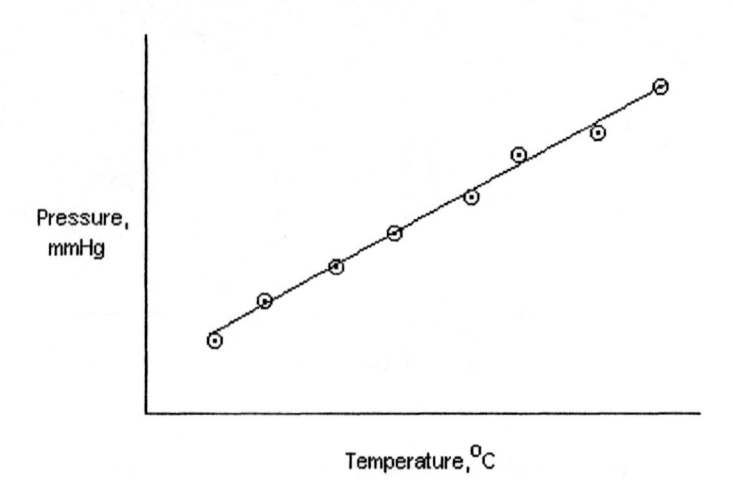

Figure 5

The slope of the line is:

$$\text{slope} = \frac{\Delta Y}{\Delta X}$$

The slope of this line would represent the change in pressure per degree change in temperature. It can be determine by choosing two points; one near each end of the line, and then drawing horizontal and vertical lines to the axes to determine the values at these intersections, as shown in **Figure 6**.

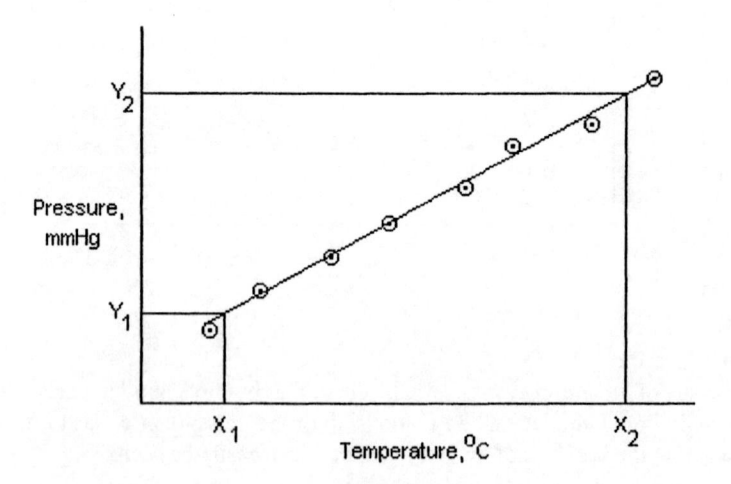

Figure 6

The slope is then:

$$\text{slope} = \frac{Y_2 - Y_1}{X_2 - X_1}$$

Name:_____   Section:_____

Experiment 3 Problems

1. When constructing a graph from experimental data, what important points should be kept in mind in order to get a well-designed graph?

   a.

   b.

   c.

   d.

   e.

   f.

Experiment 3 The Graphing of Data

2. For more than a century, except for a few years during World War II, pennies have been made of copper. After 1982, pennies were no longer made completely of copper, but rather had a zinc core with a thin copper coating. This was done since zinc was much cheaper than copper. When a series of pennies were weighed, the following masses were determined. Plot this data and draw a line that represents the average mass of the pennies before 1982. Draw a second line that represents the average mass of the pennies after 1982.

Date	Mass, g	Date	Mass, g	Date	Mass, g	Date	Mass, g
1970	2.72	1976	2.93	1982	2.48	1988	2.51
1971	2.81	1977	2.88	1983	2.32	1989	2.43
1972	3.02	1978	2.71	1984	2.59	1990	2.55
1973	2.95	1979	3.00	1985	2.41	1991	2.49
1974	2.98	1980	2.97	1986	2.47	1992	2.45
1975	2.87	1981	3.05	1987	2.61	1993	2.56

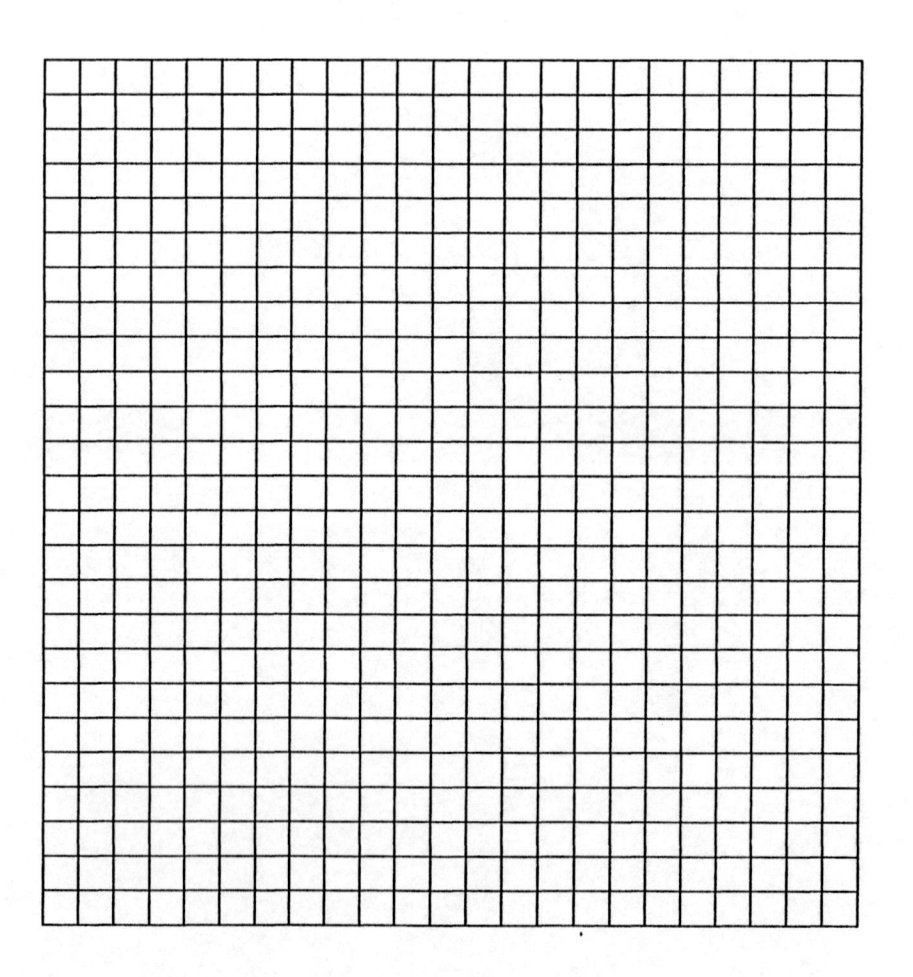

Experiment 3 The Graphing of Data

3. a. When the temperature of water is determined by using a Celsius and a Fahrenheit thermometer, the following data are obtained. Plot this data using the Celsius temperatures as the independent variable.

°C	°F	°C	°F	°C	°F
10	50	30	86	60	140
15	59	45	113	65	149
25	77	50	122	75	167

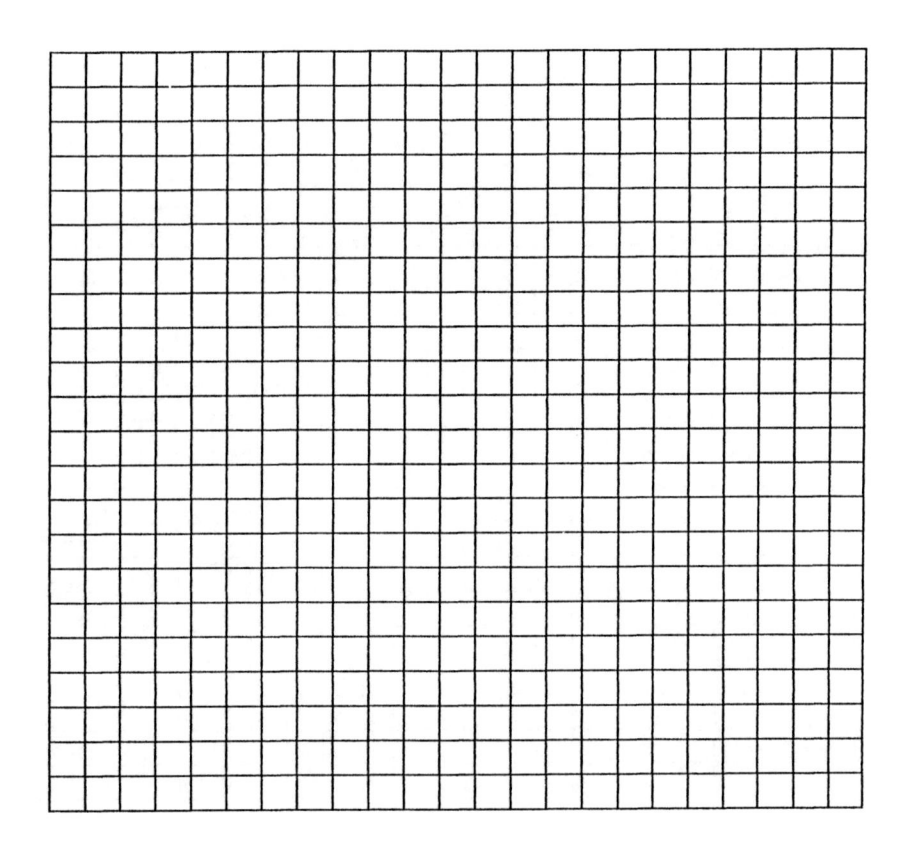

b. Body temperature is 98.6°F. From your graph, determine the corresponding temperature in °C by interpolation?

c. Extrapolate your line to determine the Fahrenheit temperature that corresponds to 0°C. What value did you get?

d. Determine the slope of the line that you have drawn. Show your work.

Experiment 3 The Graphing of Data

4. a. The boiling point of water depends on the atmospheric pressure. From the following data, make a plot of the boiling point of water versus the atmospheric pressure. Allow the pressure to be the independent variable.

Atmospheric Pressure, mmHg	Temperature, °C	Atmospheric Pressure, mmHg	Temperature, °C
750	99.63	760	100.00
752	99.70	762	100.07
754	99.78	764	100.15
756	99.85	766	100.22
758	99.93	768	100.29

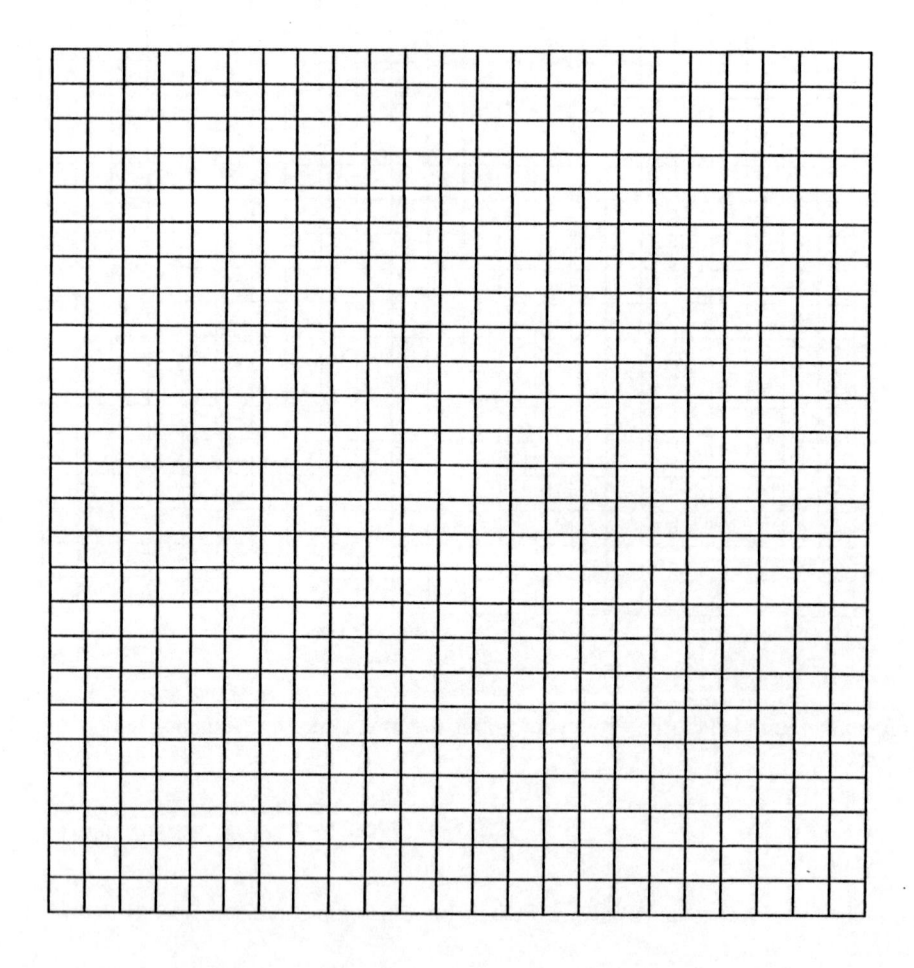

b. From the graph, determine the boiling point of water when the atmospheric pressure is 757 mmHg.

5. When the pressure of a gas is increased while its temperature is held constant, its volume decreases. This is referred to as Boyle's Law. Use the following data to plot a graph showing this change.

Pressure, mmHg	Volume, mL	Pressure, mmHg	Volume, mL
115	834	500	190
264	361	573	167
351	275	625	154
406	233	680	141
442	217	760	126

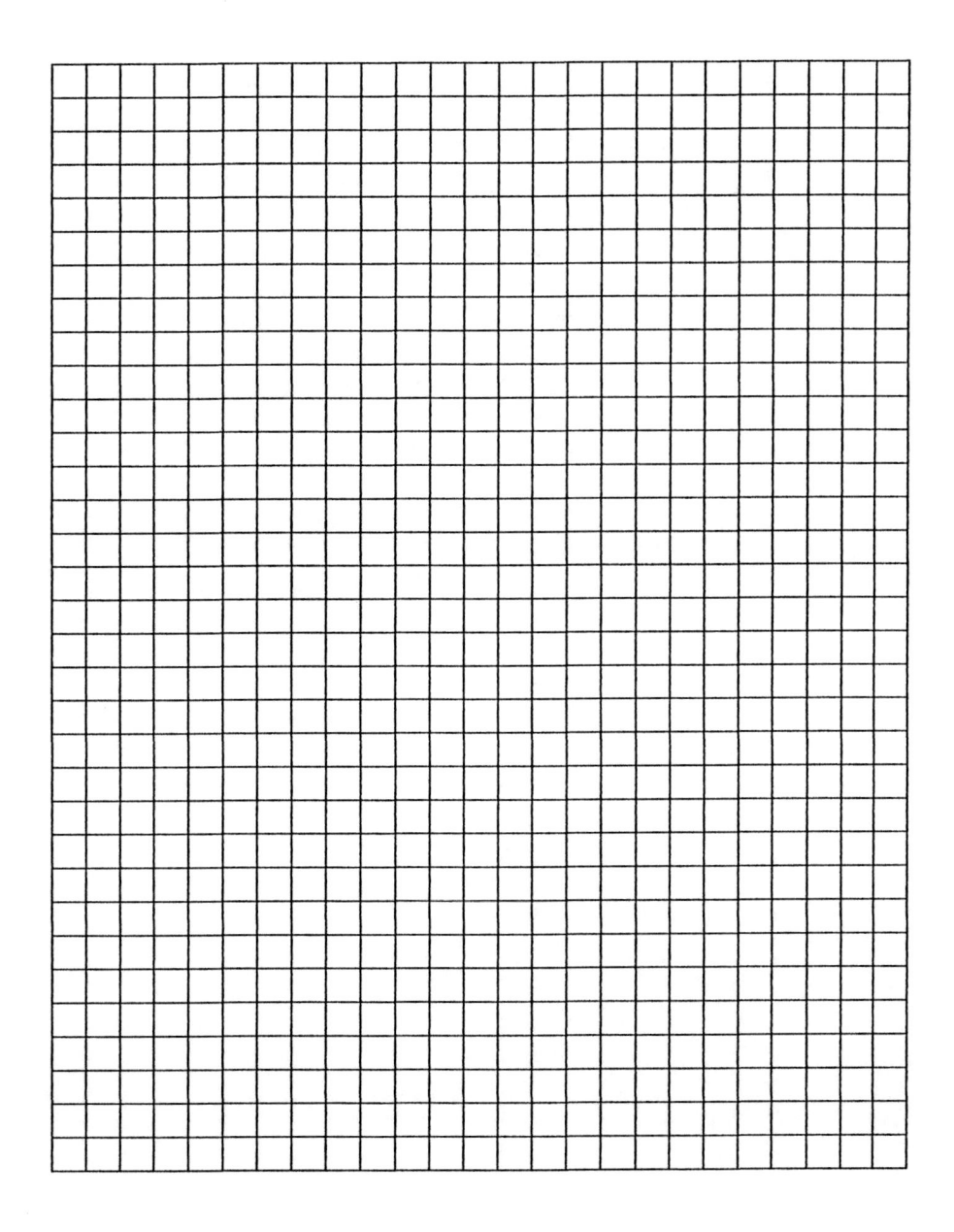

Experiment 3 The Graphing of Data

Name:_____   Section:_____

EXPERIMENT 4

Measurements of Length and Mass

I. OBJECTIVES:

A. To properly measure the dimensions of a wooden block and calculate its volume in metric units.

B. To properly determine the weights of a set of pennies by using a mechanical balance.

C. To make a plot of the mass of the pennies vs their dates.

II. DISCUSSION:

The first part of this experiment will involve measuring the dimensions of a wooden block, using a metric ruler. Obtaining the most accurate results depends on using the metric ruler to its maximum capacity. As a rule of thumb, you can always read the scale of any instrument to one place past the smallest division that is directly shown on the scale. **In other words, record all digits that are shown by the smallest line divisions on the scale plus one more digit that is estimated (sometimes termed the uncertain digit).** These digits that you record, are referred to as **significant digits** or **significant figures**.

Suppose we are measuring the length of a piece of copper wire using a metric ruler as shown below.

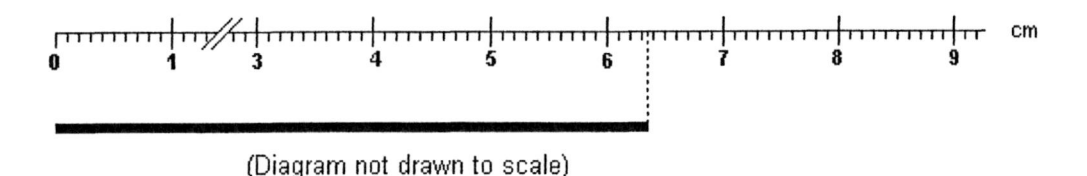

(Diagram not drawn to scale)

The length of the wire is somewhere between 6.3 and 6.4 cm. Since the scale is calibrated to the tenths place, we can estimate to one more place, i.e., the hundredths place. Thus, the length of the wire, by estimation of this last place, is 6.35 cm.

If another piece of wire went exactly to the 7.8 cm mark, the value that we write down should be such that it is as accurate as the preceding measurement. Since the first measurement could be estimated to the hundredths place, this one can also be estimated to that decimal place. We should therefore record our measurement as: 7.80 cm. While this may seem rather picky in terms of our everyday experience, it can be quite important in scientific measurements, since this value may be used in a calculation where the result has to be rounded off. The extent to which the result is rounded is determined by the number of significant figures in the measurements that we have made.

Experiment 4 Measurements of Length and Mass

After measuring the length, width, and height of the block of wood, you will calculate its volume and round off the answer to the appropriate number of significant figures.

The second part of the experiment will involve measuring the mass of a series of pennies. A graph will then be plotted of the masses versus the dates on the pennies in order to determine the average mass of a penny.

III. MATERIALS:

A lettered wooden block, metric ruler, balance, and vial of pennies.

IV. PROCEDURE:

A. The Volume of a Block of Wood:

1. Obtain a block of wood from the instructor and write down its letter in **Table 1**.

2. Using the metric ruler, measure the dimensions of the side on which the letter is written. The longest side will give you the length measurement and the other side will give you the width dimension. Make both of these measurements in the center of the block of wood. Record both of these dimensions below. Now measure the third dimension of the block, the height or thickness, and record it in the **Table**.

3. Calculate the volume of the block of wood using the following equation, and record it in the **Table**. Don't forget to round off your answer to the correct number of significant figures.

$$\text{Volume} = \text{length} \times \text{width} \times \text{height}$$

4. Convert each measurement to units of millimeters and enter the values in the Table. Again calculate the volume using units of mm for dimensions and expressing the volume in mm^3.

Table 1

Unknown Letter	Length, cm	Width, cm	Height, cm	Volume, cm^3
	Length, mm	Width, mm	Height, mm	Volume, mm^3

B. The Mass of Pennies:

1. Work with one or two partners in this experiment. Obtain a vial of pennies. After the balance is zeroed, place a penny on the pan and record the mass and the date of the penny in **Table 2**. Weigh the rest of the pennies and record the data. It is not necessary to rezero the balance before the weight of each penny.

Experiment 4 Measurements of Length and Mass

Table 2

Date	Mass, g	Date	Mass, g	Date	Mass, g

2. Plot this data on the graph below. Draw a point at the intersection of each penny's date and mass. The masses of the pennies minted before 1982 are heavier than those minted after 1982. Try to estimate the average mass of these two types of pennies from this graph. Draw a **straight, horizontal line** completely across the graph that best estimates the average mass of the pre-1982 pennies. What is the average mass of the pennies minted before 1982? _____ g Draw a **straight, horizontal line** completely across the graph that best estimates the average mass of the post-1982 pennies. What is the average mass of the pennies minted after 1982? _____ g

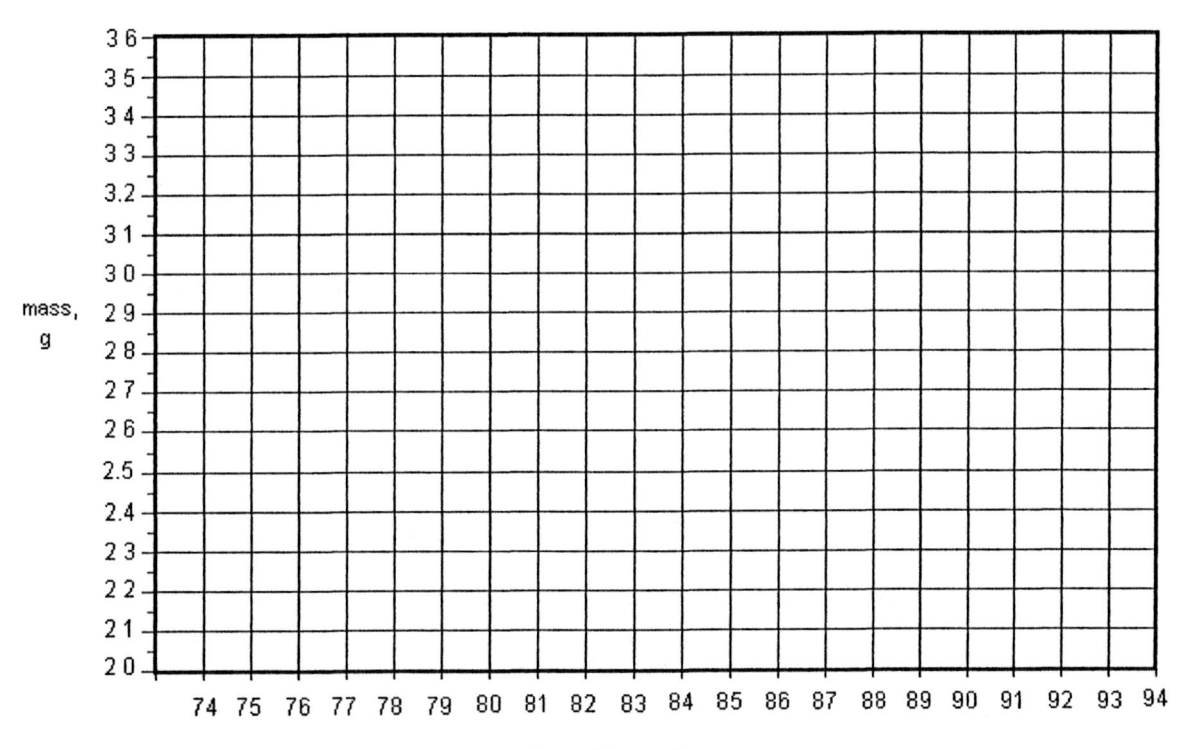

mass, g vs. Date of Penny, 19__ (y-axis: 2.0 to 3.6; x-axis: 74 to 94)

Experiment 4   Measurements of Length and Mass

Name: _____   Section: _____

Experiment 4 Problems

1. Convert your own personal height in inches into metric units of centimeters. See Appendix B for conversion factors.

2. Convert your own personal weight in pounds into metric units of grams.

3. Why do significant figures need to be considered in scientific calculations?

4. In each of the following pairs, circle the unit that is larger.

   a. dg or µg   d. g or ng

   b. dL or L   e. kg or mg

   c. nm or mm   f. µL or mL

5. The average distance from the earth to the sun is 1.50×10^8 km. What is this distance in miles?

6. A block of wood has the following dimensions: 15.75 cm × 8.66 cm × 3.10 cm
   Calculate its volume in cm^3.

Experiment 4 Measurements of Length and Mass

7. Make the following conversions and round off the answer to the correct number of significant figures. See Appendix B for conversion factors.

a. 62.5 ft to cm

b. 2.68 in to nm

c. 3.25 yd to m

d. 5.67 nm to mm

e. 4.37 cm to mm

f. 7.45 km to mm

Experiment 4   Measurements of Length and Mass

8. At the mall, you decide to try on a pair of French jeans. Naturally, the waist size of the jeans is given in centimeters. What does a waist measurement of 53 cm correspond to in inches?

9. For a pharmacist dispensing pills or capsules, it is often easier to weigh the medication to be dispensed rather than to count the individual pills. If a single antibiotic capsule weighs 0.65 g, and a pharmacist weighs out 15.6 g of capsules, how many capsules have been dispensed?

10. On your graph of the mass of the pennies, you drew a horizontal line for the pennies minted before 1982. From this line, you were able to estimate the average mass of these pre-1982 pennies. You drew another horizontal lie for the pennies minted after 1982. From this line, you were able to estimate the average mass of the post-1982 pennies.

a. Add up the masses of all the pre-1982 pennies and determine their average mass by dividing by the number of these pennies. What is the average mass of pre-1982 pennies?

b. Add up the masses of all the post-1982 pennies and determine their average mass by dividing by the number of these pennies. What is the average mass of post-1982 pennies?

c. How do these averages compare to the values you obtained by drawing straight lines on your graph?

d. Why do you think there is a difference in the mass of pre- versus post-1982 pennies?

Name:_____   Section:_____

EXPERIMENT 5

The Measurement of Density

I. OBJECTIVES:

A. To measure the mass and volume of several pieces of aluminum.

B. To properly construct a graph by plotting mass <u>vs</u> volume.

C. To determine the density of the aluminum from the slope of the graph.

II. DISCUSSION:

The density of a substance is defined as its mass per unit volume, i.e.,

$$d = \frac{m}{V}$$

Thus, in order to determine the density of a substance, its mass and volume must be measured.

In this experiment you will first determine the mass of aluminum slugs by weighing them on a balance. Since the shapes of the pieces of aluminum are irregular and cannot be calculated from a geometry formula, they must be measured experimentally. These volumes will be determined by the method of **water displacement**.

In general, this method consists of putting some water in a container, recording the water volume, adding the piece of aluminum, and recording the new volume. By taking the difference of the volumes, the volume of the aluminum can be obtained. Beakers and flasks are graduated in only approximate values; usually ± 10 %. To make a more accurate volume measurement, a graduated cylinder is used.

Liquids in a graduated cylinder form a curved surface called the **meniscus**. In order to read the volume properly and consistently, the graduated cylinder is held vertically and the bottom of the meniscus is read. In order to avoid **parallax** error (different values of a measurement due to different positions of the observer), the bottom of the meniscus must be at eye level. If it is held above or below eye level, different volume values are obtained (try it and see).

Experiment 5 The Measurement of Density

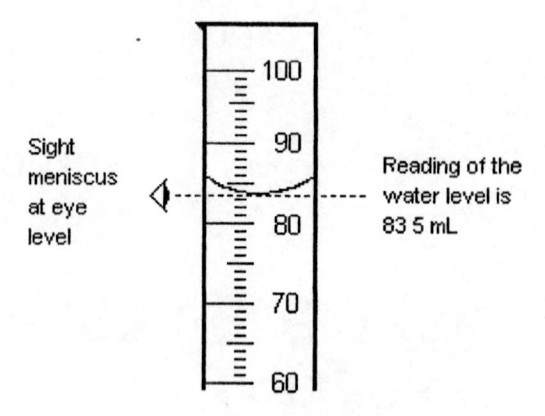

[Figure: graduated cylinder showing meniscus at eye level, reading of the water level is 83.5 mL]

After measuring the masses and volumes of a series of aluminum pieces, the density will be determined by making a plot of mass (on the Y-axis) vs volume (on the X-axis). The data points should fall on approximately a straight line. A best fit line is drawn. Experimental error may result in some points being slightly off the best fit line. The slope of the line, which also will be the density of the aluminum, will be determined.

The slope of a straight line is defined as:

$$\text{slope} = \frac{Y_2 - Y_1}{X_2 - X_1}$$

where Y_2 and X_2 are the larger mass and volume values, and Y_1 and X_1 are the smaller mass and volume values.

Keep in mind the following when you draw your graph:

1. The graph should be neatly titled, labeled, and drawn.
2. Use a ruler when drawing lines.
3. Each axis should be labeled appropriately, with the unit of the variable following a comma.
4. The scales for each axis should be adjusted so that the range of data spans most of the axis.
5. It is not necessary that the scales begin with zero.
6. When constructing a scale, the subdivisions between major divisions should allow for easy interpretation. For example, if there are 10, 5, or 3 subdivisions and then the subdivisions are 0.1, or 0.2 or 0.3 units in a division. The last one is inconvenient.

[Figure: three scale examples showing 0.1 divisions, 0.2 divisions, and 0.333 divisions between marks 1 and 2]

7. The size of the divisions and subdivisions on one scale need not be the same size as the divisions and subdivisions on the other scale.
8. When plotting data, place a data point on the graph and then draw a circle around it so that it is more visible.

Experiment 5 The Measurement of Density

III. MATERIALS:

A vial with a set of aluminum pieces, 100 mL graduated cylinder, electronic balance, and a sheet of graph paper.

IV. PROCEDURE:

A. Mass Determination:

1. Work in pairs on this experiment. Obtain a container of aluminum pieces from the instructor.

2. Remove the pieces of aluminum from the container and weigh each one, recording the weights in **Table 1**.

B. Volume Determination:

1. Add approximately 60 mL of water to the 100 mL graduated cylinder. Read the water level in the cylinder and record this water volume in the Table as the first Initial Volume.

2. Add one of the aluminum pieces to the graduated cylinder (slide the piece down the side of the graduated cylinder so you do not break its bottom. Tap the cylinder a few times to dislodge any air bubbles that might have clung to the aluminum. Measure and record this Final Volume in the Table.

3. Record this same volume in the Table as the Initial Volume for the second piece of aluminum.

4. Add the second piece of aluminum to the cylinder and as before measure and record the Final Volume.

5. Repeat this procedure for the remaining pieces of aluminum.

6. Subtract the Initial Volumes from the Final Volumes of each piece of aluminum in order to obtain the Volume of the pieces of aluminum. Record these values in the Table.

Table 1

Piece of Aluminum	Mass, g	Initial Volume, mL	Final Volume, mL	Volume, mL
1				
2				
3				
4				
5				

Experiment 5 The Measurement of Density

C. Constructing the Graph:

1. Make a plot of your mass vs volume data on the graph provided on the next page. The division of units on each axis should be selected so that the data extends over most of the length of the axis, rather than being cramped into one tiny section of the graph. Additionally, the distances between the units should be such that it is easy to estimate a value of mass or volume off the graph. In other words, each division on an axis should represent 1,2, or 5 units (or multiples or submultiples of these). Use the X-axis for volume in units of mL and the Y-axis for mass in units of grams. Title the graph, **The Measurement of the Density of Aluminum** and label the axes appropriately.

2. Using a ruler, draw a straight line that **best fits** the plotted data points. Do not draw a line connecting the "dots." The best fit line can most easily be determined by using your ruler and sighting along the data points from an extreme oblique angle to the surface of the paper (the squint test).

D. Determining the Density:

Calculate the slope of the graphed line by using the following equation:

$$\text{slope} = \text{density} = \frac{Y_2 - Y_1}{X_2 - X_1}$$

where Y_2 and X_2 are the values of some point on the line near its right end, and Y_1 and X_1 are the values of some point on the line near its left end. **Do not use any of your experimental data points.** Show this calculation of the density of the aluminum in the space below.

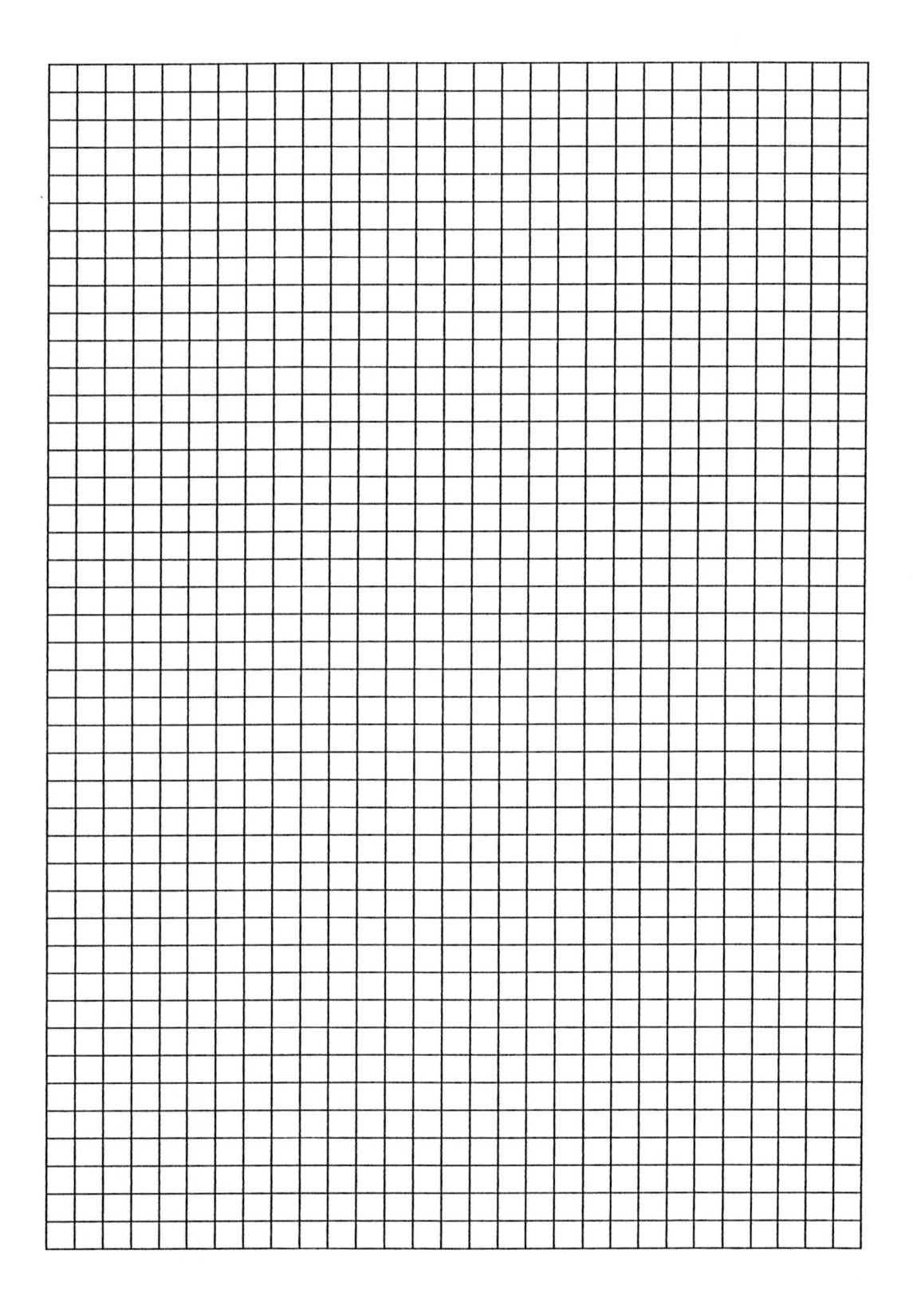

Experiment 5 The Measurement of Density

Experiment 5   The Measurement of Density

Name:_____   Section:_____

Experiment 5 Problems

1. What is the definition of density?

2. What happens to the density of a substance as its temperature increases?

Explain why?

3. What is parallax?

4. A piece of silver (density = 10.5 g/cm^3) was found to weigh 55.2 g. What is its volume?

5. Zinc has a density of 7.14 g/cm^3. A rectangular block of zinc was measured and found to have a length of 15.67 cm, a width of 9.21 cm, and a thickness of 2.66 cm. What is the mass of this zinc block?

Experiment 5   The Measurement of Density

6. The density of silver is 10.5 g/cm^3 at 20.°C. If 46.3 g of pure silver pellets are added to a graduated cylinder containing 15.4 mL of water, what will be the new water volume in the cylinder?

7. At a local pawn shop a student finds a medallion that the pawn shop owner insists is pure platinum. However, the student suspects that the medallion may actually be silver, and thus much less valuable. The student buys the medallion only after the shop owner agrees to refund the price if the medallion is returned within two days. The student, a chemistry major, then takes the medallion to her lab and measures the density as follows. She first weighs the medallion and finds its mass to be 55.64 g. She then places some water in a graduated cylinder and reads the volume as 62.8 mL. Next she drops the medallion into the cylinder and reads the new volume as 65.4 mL. Is the medallion platinum (density = 21.4 g/cm^3) or silver (density = 10.5 g/cm^3)?

8. A solid metal sphere has a volume of 6.33 cm^3. The mass of the sphere is 45.00 g. Find the density of the metal sphere.

9. If you were asked to measure 3.7 mL of a liquid, what would be the best piece of glassware to use?

a. 100 mL graduated cylinder
b. 10 mL graduated cylinder
c. 100 mL beaker
d. 10 mL beaker
e. 125 mL Erlenmeyer flask

Explain why.

Experiment 5   The Measurement of Density

10. A student obtained the following data for the mass and volume of an unknown substance.

Sample #	Mass, g	Volume, mL
1	15.37	1.96
2	18.91	2.40
3	20.33	2.57
4	25.56	3.01
5	31.14	3.98

Plot this data, on the graph below, and determine the best value for the density of the substance, just as you did with your data for aluminum. Remember to show your calculation of the slope.

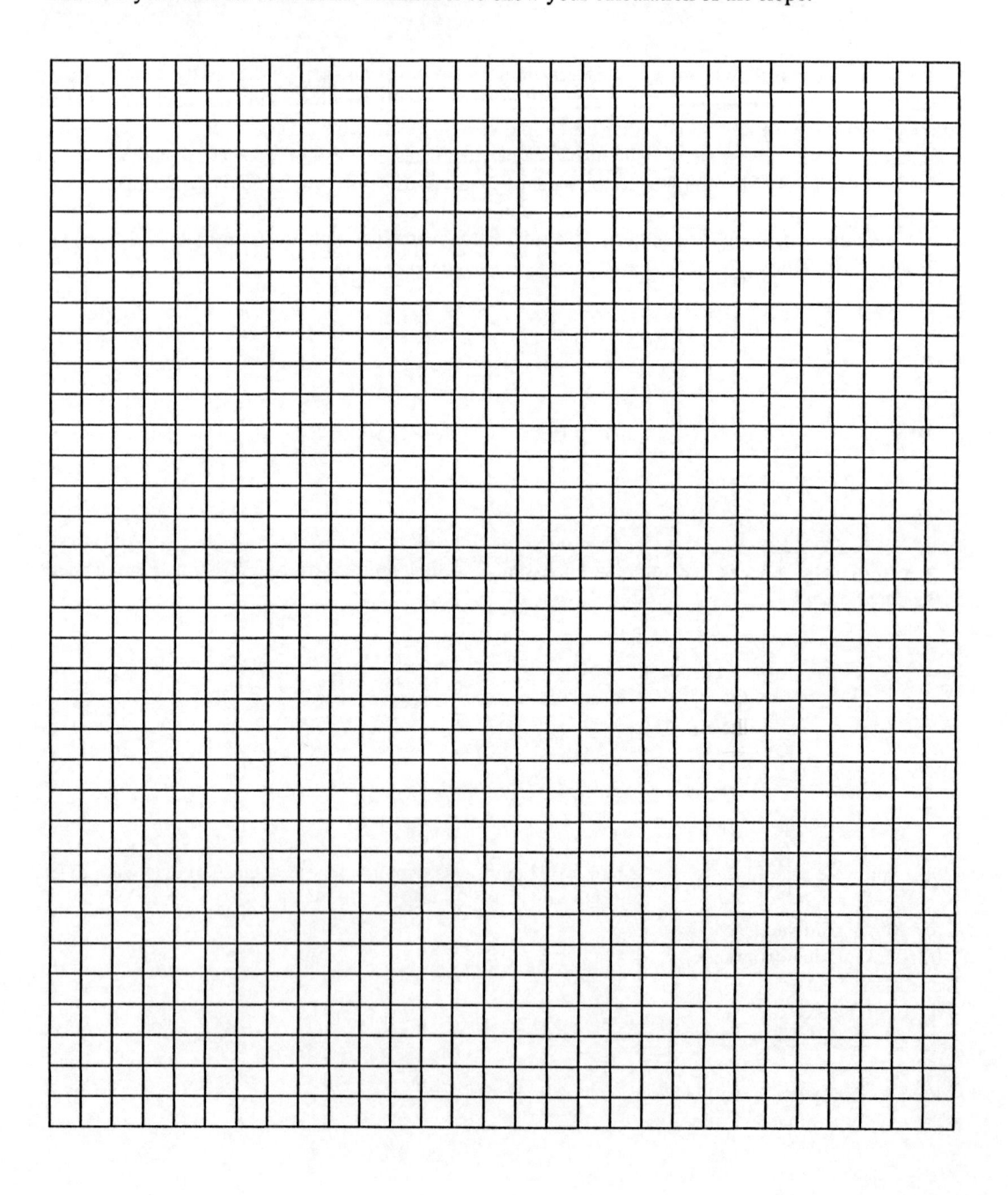

Name:_____   Section:_____

EXPERIMENT 6

Periodic Properties of Elements

I. OBJECTIVES:

A. To examine some physical and chemical properties of Group I and II metals.

B. To examine the elements in Period 3.

C. To test the properties of the Group II oxides and compare them to the properties of a Group V oxide.

D. To prepare the halogens and test some of their properties.

II. DISCUSSION:

The arrangement of the elements in the Periodic Table has proven extremely useful to chemists (and to students who are studying chemistry). The modern Periodic Table is based on the contributions of a Russian scientist, Dmitri Mendeleev, and a German scientist, Julius Lothar Meyer. In 1869 they independently developed tables in which the elements were arranged in order of increasing atomic mass (or atomic volumes). In addition, elements with similar chemical properties were placed in the same vertical columns (Groups). This arrangement had a number of gaps in it since less than 70 elements were known at the time. Mendeleev predicted that these gaps were due to the existence of yet undiscovered elements. For example, gaps existed between zinc and arsenic. Since one of these gaps was underneath the element silicon, Mendeleev called this missing element "ekasilicon." By 1886, this gap was filled by the discovery of the element germanium. Its properties very closely matched those that had been predicted by Mendeleev.

Some of the elements seemed to be out of order on this early Periodic Table. While iodine has chemical properties similar to those of bromine and chlorine, its atomic mass placed it in the preceding Group. In 1913, Harry Moseley concluded that the elements should be arranged in order of increasing atomic number. This placed iodine in the correct Group, in addition to those other elements that were out of order. This led to the current **Periodic Law**: when the elements are arranged in order of increasing atomic number, their properties vary periodically.

In this experiment, you will test the properties of some elements in selected Groups. You will also compare elements in a Period in order to see differences in their properties.

In the first part of the experiment, you (or your instructor) will test the properties of the Alkali Metals (Group IA). When lithium is added to water, a rapid reaction occurs, generating hydrogen gas:

$$2\,Li \;+\; 2\,H_2O \;\rightarrow\; 2\,LiOH \;+\; H_2$$

Similar reactions occur with sodium and potassium. These reactions will be compared to those that Group IIA metals undergo with water.

Most of the elements are metals. If elements are divided into metals and nonmetals, only 21 of them are nonmetals. These nonmetals are found on the right side of the Periodic Table, and include familiar substances like carbon, oxygen, nitrogen, sulfur, iodine, and helium. You will examine the elements of Period 3 from sodium to chlorine. By visual inspection, you will observe how their physical characteristics vary in going to the right in this Period.

When the elements of a group combine with some other element, the resulting compounds may exhibit similar chemical properties. You will test the reaction of the oxides of some Group IIA metals with water. The general reaction may be illustrated with calcium oxide.

$$CaO + H_2O \rightarrow Ca(OH)_2$$

A property of this product, calcium hydroxide, will be compared to that of the products of strontium oxide and barium oxide with water. The reaction of these metal oxides will be compared to that of nonmetal oxides. An oxide of phosphorus, P_4O_{10}, will be added to water. The reaction that occurs is:

$$P_4O_{10} + 6 H_2O \rightarrow 4 H_3PO_4$$

The product of the resulting reaction will be tested and compared to the earlier tests of the metal oxide products.

An oxide of sulfur, sulfur dioxide, will be prepared by the reaction of sodium sulfite and hydrochloric acid:

$$Na_2SO_3 + 2 HCl \rightarrow 2 NaCl + H_2O + SO_2$$

The resulting SO_2 will be bubbled into distilled water, in which the following reaction will occur.

$$H_2O + SO_2 \rightarrow H_2SO_3$$

This solution will be tested and compared to the result obtained with H_3PO_4.

The Group VIIA elements (halogens) will be examined and their properties noted. Water solutions of the halogens will be shaken with mineral oil. The halogens will preferentially dissolve in one of the liquids and the color of the solution will be noted.

The halogens will also be reacted with the halide ions in order to determine relative reactivities. For example, the following reaction occurs when sodium iodide is added to bromine water:

$$2 NaI + Br_2 \rightarrow 2 NaBr + I_2$$

You will test for the presence of the iodine in the product.

The halide ions will be tested for their reaction with silver ion. For example, when sodium chloride solution is mixed with silver nitrate solution, a precipitate is observed.

$$NaCl(aq) + AgNO_3(aq) \rightarrow NaNO_3(aq) + AgCl(s)$$

You will compare this reaction to that involving sodium bromide and sodium iodide. The silver halide products will then be tested for their solubility in ammonia water.

III. MATERIALS:

Four 250 mL beakers, knife, forceps, red litmus, blue litmus, six 6" test tubes, lithium, sodium, potassium, magnesium ribbon, calcium turnings, aluminum, silicon, phosphorus, sulfur, chlorine, calcium oxide, strontium oxide, barium oxide, tetraphosphorus decaoxide, sodium sulfite, sodium hypochlorite solution, hexane, mineral oil, 6 M hydrochloric acid, saturated sodium bromide solution, concentrated sulfuric acid, 20% sodium iodide solution, chlorine water, 10% sodium bromide solution, 5% sodium iodide solution, bromine water, 5% sodium chloride solution, iodine water, 2% silver nitrate solution, 6 M ammonia water, 15 M ammonia.

IV. PROCEDURE:

A. The Reaction of Alkali Metals with Water:

Your instructor may choose to do this experiment as a demonstration. Record all your observations in **Table 1**.

1. To each of three 250 mL beakers, add about 100 mL of water.

2. Cut a small piece of lithium **(Li)** (about 1/4 cm on a side) with a knife, while holding it with forceps. Observe its characteristics and record your observations in **Table 1**. The alkali metals are stored under mineral oil or some relatively inert liquid. Wipe off the excess oil with a paper towel.

3. Using the forceps, drop the lithium metal into one of the beakers. What do you observe occurring?

4. Repeat the above procedure using sodium **(Na)** and potassium **(K)**.

5. Using red and blue litmus paper, test all three of the above solutions.

6. Pour the solutions into the sink and wash the beakers with soap and water.

Table 1

	Lithium	**Sodium**	**Potassium**
Appearance			
Ease of Cutting			
Reaction with Water			
Red Litmus Test			
Blue Litmus Test			

Experiment 6 Periodic Properties of Elements

B. The Reaction of Alkaline Earth Metals with Water:

1. To each of two 250 mL beakers, add about 100 mL of water. Obtain a small piece of magnesium (**Mg**), ribbon and observe its properties. Record these in **Table 2**.

2. Add the magnesium to the water and record and changes that you observe.

3. Use a forceps to obtain and hold a piece of calcium (**Ca**). Observe its appearance, and record these observations in the Table.

4. Add the calcium to the second beaker of water and observe any changes that may occur. Record your observations.

5. Test each of the beakers of water with red and blue litmus.

6. Add 10 drops of 6 M hydrochloric acid (**HCl**) to each beaker. Record your observations.

6. Pour the liquids into the sink and place the unreacted metals in the container designated by the instructor. Wash the beakers with soap and water.

Table 2

	Magnesium	Calcium
Appearance and Properties		
Reaction with Water		
Red Litmus Test		
Blue Litmus Test		
Reaction with HCl		

Experiment 6 Periodic Properties of Elements

C. The Physical Properties of the Elements of Period 3:

1. Obtain containers of the elements found in Period 3 of the Periodic Table, i.e., sodium **(Na)**, magnesium **(Mg)**, aluminum **(Al)**, silicon **(Si)**, phosphorus **(P)**, sulfur **(S)**, and chlorine **(Cl)**.

2. Observe these elements and record your observations in **Table 3**. Return the containers when you have finished with your observations.

Table 3

	Appearance
Sodium	
Magnesium	
Aluminum	
Silicon	
Phosphorus	
Sulfur	
Chlorine	

D. The Reaction of Metal Oxides and Nonmetal Oxides with Water:

1. To each of four 250 mL beakers, add about 100 mL of water.

2. Obtain and observe the properties of the following Group IIA oxides: calcium oxide **(CaO)**, strontium oxide **(SrO)**, and barium oxide **(BaO)**. Record these observations in **Table 4**.

3. To the first beaker, add about 1 g of calcium oxide, and record what you observe. To the second beaker, add about 1 g of strontium oxide, and to the third beaker, add about 1 g of barium oxide. Again, record your observations.

4. Obtain a sample of tetraphosphorus decaoxide **(P_4O_{10})** from the instructor.

Caution: P_4O_{10} is corrosive. Avoid any contact with your skin.

Observe and record the appearance of this oxide. To the fourth beaker of water, carefully add the tetraphosphorus decaoxide and record what you see happening.

Experiment 6 Periodic Properties of Elements

5. Test the solutions in all four beakers with red and blue litmus paper.

6. Dispose of these four solutions as directed by the instructor and then wash the beakers with soap and water.

7. Assemble a 6" test tube fitted with a one-holed stopper containing a short glass tube connected to a piece of rubber tubing. At the end of the rubber tubing attach a piece of glass tube about 6" long.

8. Fill a second 6" test tube about half full with water (see **Figure 1**).

9. **Perform the following reaction in the hood.** To the first test tube, add about 1 g of sodium sulfite **(Na_2SO_3)** and then about 2 mL of 6 M hydrochloric acid **(HCl)**. As the reaction occurs, a gas is evolved. Place the glass tube into the water of the second test tube. As the gas bubbles into the water, some of it will dissolve and react. After about a minute, remove the tube and test the water with red and blue litmus paper. Record your observations in the Table.

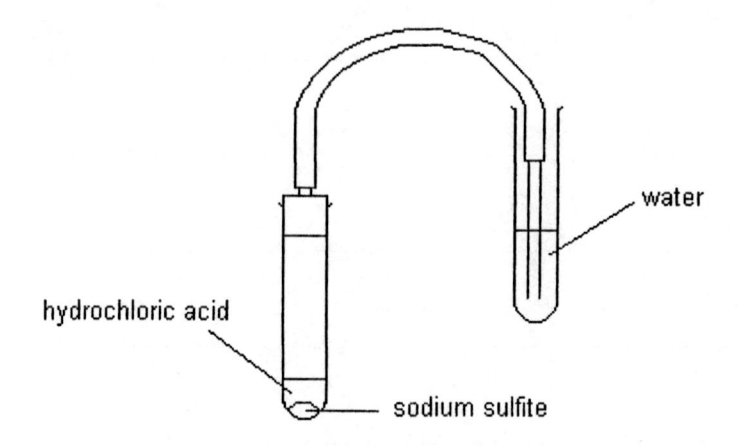

Figure 1

Table 4

	CaO	SrO	BaO	P_4O_{10}	SO_2
Appearance					
Reaction with Water					
Red Litmus Test					
Blue Litmus Test					

Experiment 6 Periodic Properties of Elements

E. The Preparation of the Halogens:

1. **Perform the following tests in the hood.** To a 6" test tube, add about 2 mL of sodium hypochlorite (**NaOCl**) solution and about 1 mL of hexane. The two liquids are not miscible in each other. Which is the top layer?

Top layer: _____

2. To this mixture, slowly add dropwise about 1 mL of 6 M hydrochloric acid (**HCl**) to the test tube. As the chlorine is produced, it will dissolve in the hexane. Note the color of the chlorine that is dissolved in the hexane. Record this in **Table 5**.

3. To a second 6" test tube, add about 2 mL of a saturated sodium bromide (**NaBr**) solution and about 1 mL of hexane. To this mixture, slowly add dropwise about 3 mL of concentrated sulfuric acid (**H_2SO_4**). **Caution: concentrated sulfuric acid is extremely corrosive. Wash your skin thoroughly if you come in contact with any acid.** As you add the sulfuric acid, agitate the test tube in order to mix the reactants. As the bromine is produced, it will dissolve in the hexane. What is the color of the bromine? Record your observations in the Table.

4. To a third 6" test tube, add about 2 mL of a 20% sodium iodide (**NaI**) solution and about 1 mL of hexane. Slowly add dropwise 1-2 mL of conc. sulfuric acid, agitate the test tube gently, and observe the color of the iodine as it is produced and dissolves in the hexane. Again record your results.

5. Dispose of the contents of the test tubes as directed by the instructor, and wash the tubes thoroughly with soap and water.

Table 5

	Chlorine	Bromine	Iodine
Color of the Hexane Layer			

F. The Reaction of Halogens with Halide Ions:

1. Obtain six 6" test tubes and label them 1-6.

2. Add about 2 mL of chlorine (**Cl_2**) water to test tubes 1 and 2. To each of these test tubes, add about 1 mL of hexane. Agitate the tubes for about 10 seconds. Observe the color of the hexane layer and record this in **Table 6**.

3. To test tube 1, add about 1 mL of 10% sodium bromide (**NaBr**) solution. Stopper the test tube and agitate for about 10 seconds. Observe and record the color of the hexane layer.

4. To test tube 2, add about 1 mL of 5% sodium iodide (**NaI**) solution. Stopper the test tube and shake for about 10 seconds. Record the color of the hexane.

5. To each of test tubes 3 and 4, add about 2 mL of bromine (**Br_2**) water and about 1 mL of hexane. Agitate the tubes for about 10 seconds. Record the color of the hexane.

Experiment 6 Periodic Properties of Elements

6. To test tube 3, add about 1 mL of 5% sodium chloride **(NaCl)** solution. To test tube 4, add about 1 mL of 5 % sodium iodide **(NaI)** solution. Agitate both test tubes and record the resulting color of the hexane.

7. To each of test tubes 5 and 6, add about 2 mL of iodine water and about 1 mL of hexane. Agitate the tubes for about 10 seconds and record the colors of the hexane.

8. To test tube 5, add about 1 mL of 5% sodium chloride **(NaCl)** solution. To test tube 6, add about 1 mL of 10% sodium bromide **(NaBr)** solution. Agitate the tubes for about 10 seconds. Record the colors that you observe in the hexane.

9. Dispose of the test tubes contents as directed by the instructor, and then wash them thoroughly with soap and water.

Table 6

	Color of the Hexane Before Reaction	Color of the Hexane After Shaking with the Halides		
		NaCl	NaBr	NaI
Chlorine Water				
Bromine Water				
Iodine Water				

G. The Reaction of Halides with Silver Nitrate:

1. Label three 6" test tubes 1-3.

2. To test tube 1, add 2 mL of 0.1 M sodium chloride **(NaCl)** solution and two drops of a 2% silver nitrate **(AgNO₃)** solution. Observe what occurs, and record these observations in **Table 7**.

3. To test tube 2, add 2 mL of 0.1 M sodium bromide **(NaBr)** solution. To test tube 3, add 2 mL of 0.1 M sodium iodide **(NaI)** solution. To both tubes, add two drops of the 2 % silver nitrate solution. Record the changes that you see in the Table.

4. After the precipitates have settled, decant off the liquid in all three test tubes. To each of the test tubes, add 6 M ammonia **(NH₃)** solution dropwise. Agitate the test tubes during the additions. Record any changes that may occur during the ammonia water addition.

5. If no change has occurred after about 2 mL of the ammonia water has been added, discontinue the addition, allow the solid to settle, and decant off the liquid.

6. To the test tubes containing solid, add 15 M ammonia dropwise, and agitate the test tubes. Caution: Concentrated ammonia has an extremely strong, corrosive odor. Again, discontinue the addition after about 2 mL of the ammonia water has been added. Observe any changes that may have occurred in the tubes and record your results in the Table.

7. Dispose of the contents of the test tubes in the container provided by the instructor. Wash the tubes thoroughly with soap and water, and then return all equipment and chemicals to the designated areas.

Table 7

	NaCl	NaBr	NaI
Reaction with $AgNO_3$			
Solubility of Precipitate in 6 M Ammonia Water			
Solubility of Precipitate in 15 M Ammonia Water			

Experiment 6 Periodic Properties of Elements

Name:_____   Section:_____

Experiment 6 Problems

1. a. In the following pairs, which is the more reactive metal?

    1) sodium or magnesium

    2) lithium or potassium

    3) calcium or potassium

   b. What evidence do you have to support your choices?

2. Write the equation representing the reaction of potassium with water.

3. Write the equation representing the reaction of calcium with water.

4. When the product of the reaction of the alkali metals with water was tested with litmus paper you observed a change in the color of the litmus. What do you think caused this color change?

5. a. If a metal oxide reacts with water, what type of product, in general, is produced?

   b. If a nonmetal oxide reacts with water, what type of product, in general, is produced?

6. a. Which has a greater attraction for electrons: chlorine, bromine or iodine?

   b. What experimental evidence do you have to support this choice?

7. An unknown compound containing halide ion gave a precipitate when treated with silver nitrate. The precipitate did not dissolve in dilute ammonia water. It did however, dissolve in concentrated ammonia. What halide did the compound contain?

8. Why do the elements within a Group exhibit similar chemical properties?

9. a. Are halogens more soluble in water or hexane?

   b. What is the reason for your choice, based on the experiment you did involving these substances?

10. Predict what would happen if NaAt was added to bromine water.

11. How could you determine that the gas produced when an alkali metal reacts with water is hydrogen?

12. What is the collective name given to the elements in:

   a. Group IA

   b. Group IIA

   c. Group VIIA

   d. Group VIIIA

   e. the B Groups

13. If the Periodic Table was arranged by increasing Atomic Mass, several pairs of elements would be out of order and not exhibit the periodic recurrence of chemical properties for the Group in which they would be found. What pairs of elements would be out of order in the main body of the Periodic Table (ignore the Lanthanide and Actinide series)?

14. When you examine the elements in Period 3, what happens to metallic character as you go from left to right on the Periodic Table?

Name:_____   Section:_____

EXPERIMENT 7

Solubility and Miscibility

I. OBJECTIVES:

A. To become familiar with the components of a solution.

B. To examine some of the terms and concepts associated with solutions.

C. To investigate solute and solvent interaction.

D. To learn to make inferences about the nature of substances, based on their solubility or miscibility.

II. DISCUSSION:

In many chemical reactions, both in the laboratory and in the world around us, one or more of the reactants are present in a solution, i.e., they are dissolved in some fluid, such as water.

A solution is a homogeneous mixture, and may be of variable concentration. An example is salt dissolved in water, where the amount of salt dissolved in a given amount of water can vary from one solution to another. Our ocean is about 3 % salt while the Great Salt Lake in Utah is about 20 % salt.

We refer to the fluid component of a solution as the **solvent**. Its physical state does not change when the solution is formed. All the other components which are dissolved in the solvent are called **solutes**. These solutes may be solids, liquids, or gases.

Solubility is the property of one substance to dissolve in another substance. For example, salt has the ability to dissolve in water. Therefore, it is said to be soluble in water.

Miscibility is used to describe the solubility of one liquid in another. In this case, the solute and solvent are both liquids, but the one present in larger quantity is referred to as the solvent. If one liquid dissolves well in another, we say that the two liquids are **miscible**. If a liquid does not dissolve well in another, the two liquids are then termed **immiscible**. Miscible liquids thus form homogeneous mixtures, while immiscible liquids form heterogeneous mixtures in which the layers of the two liquids can be seen, such as oil and vinegar.

There are a number of different kinds of solutions: gases in gases (air), liquids in liquids (gasoline), solids in solids (metal alloys), gases in liquids (carbon dioxide in water), and solids in liquids (sugar water). In this experiment, you will test the solubilities of solids and the miscibility of liquids in other liquids. From your observations you will be able to draw some theoretical generalizations regarding solubilities.

Experiment 7 Solubility and Miscibility

III. MATERIALS:

A test tube rack containing 9 small test tubes and three rubber stoppers, spatula, hexane, ethanol, sodium chloride, oxalic acid, paraffin, cooking oil, dichloromethane, iodine/potassium iodide solution.

IV. PROCEDURE:

A. Solids in Liquids:

1. **Caution! Hexane and ethanol are very volatile and very flammable liquids. While working with them, absolutely no Bunsen burner flames should be in the nearby vicinity.**

2. Arrange nine clean small test tubes in a test tube rack.

3. Fill the first three test tubes about 1/4 full with water (H_2O). Label the tubes 1, 2, and 3.

4. Fill the second three test tubes about 1/4 full with ethanol (C_2H_5OH). Label these tubes 4, 5, and 6.

5. Fill the last three test tubes about 1/4 full with hexane (C_6H_{14}). It is especially important that these test tubes are **dry**. Label these tubes 7, 8, and 9.

6. Put about 0.1 g of sodium chloride (**NaCl**) in test tubes 1, 4, and 7. Stopper the test tubes and shake vigorously for about half a minute. Record whether or not the solute dissolved by putting an **S** (soluble) or **I** (insoluble) in **Table 1**. Dry the stoppers with a paper towel before you go to the next step.

7. Put about 0.1 g of oxalic acid ($H_2C_2O_4$) in test tubes 2, 5, and 8. Stopper the test tubes and shake vigorously for about half a minute. Record your observations as before. Once again, dry your stoppers before proceeding.

8. Put about 0.1 g of paraffin in test tubes 3, 6, and 9. Stopper the test tubes and shake vigorously for about half a minute. Record your observations as before.

9. Dispose of the contents of test tubes 1 through 6 down the drain and flush them down with plenty of water. Dispose of the contents of test tubes 7, 8, and 9 in the waste disposal container provided.

10. Wash all the test tubes and stoppers with soapy water and rinse them with tap water.

Table 1

Solvent	Solute		
	NaCl	$H_2C_2O_4$	Paraffin
Water	1.	2.	3.
Ethanol	4.	5.	6.
Hexane	7.	8.	9.

B. Miscibility of Liquids:

1. Add water, ethanol, and hexane to three different test tubes until each is about ¼ full. Make sure that you have dried out the tube that will contain the hexane. Add about 1 mL of cooking oil to each of the test tubes. Stopper each test tube and tilt it back and forth gently to mix the contents. Record your observations in **Table 2**.

Experiment 7 Solubility and Miscibility

Table 2

Solvent	Observations
Water	
Ethanol	
Hexane	

2. Dispose of the hexane solution in the waste disposal container provided by the instructor. Wash all three of the test tubes with soap and water.

C. The Relative Solubility of a Solute in Two Solvents:

1. To a small test tube add about 2 mL of dichloromethane **(CH_2Cl_2)** and about 4 mL of water. Observe the relative positions of each liquid, by noting the volumes of the two layers. Stopper the tube, shake for 5 seconds and allow the liquids to separate. What do you see?

Observations: _____

2. Now, **without disturbing the test tube**, add 2 drops of an iodine/potassium iodide solution to the test tube and note the color of each layer and their intensities.

Color of the aqueous layer: _____

Color of the CH_2Cl_2 layer: _____

3. Insert the stopper and shake the test tube gently for about 20 seconds. Allow the liquids to separate and again note the color of each layer.

Color of the aqueous layer: _____

Color of the CH_2Cl_2 layer: _____

4. When the iodine is shaken with the two solvents, it partitions itself between them. Based on the relative **intensity** of the colors of the two layers, in which solvent is the iodine more soluble?
_____

5. Dispose of the contents of the test tube in the waste container provided and wash the test tube and stopper with soap and water.

Experiment 7   Solubility and Miscibility

Name:_____   Section:_____

Experiment 7 Problems

1. Explain what is meant by the phrase "like dissolves like".

2. Using the solubility results obtained in this experiment, and knowing that NaCl is ionic and water is very polar, make inferences about the nature **(polar or nonpolar)** of each of the following solutes and solvents.

   Solutes: Oxalic acid   Paraffin

   Solvents: Ethanol   Hexane

3. Make an inference about the nature **(polar or nonpolar)** of cooking oil. Cite experimental evidence for your answer.

4. Explain the difference between the terms miscible and soluble.

5. In each of the following pairs, which liquid has a greater density? What experimental evidence supports your answer?

   a. dichloromethane or water

   a. water or cooking oil

6. Would you classify the dichloromethane-water mixture as miscible or immiscible? Why?

7. In the iodine experiment you worked with an iodine/potassium iodide solution which was a reddish-brown in color. This solution was a convenient source of iodine for this solubility test. Pure iodine has a

dark purplish color. When dissolved in a nonpolar solvent, iodine produces a purple solution. When the iodine solution was added to the CH_2Cl_2/H_2O mixture and the test tube was shaken, into which solvent did the iodine dissolve? Explain why this occurred.

8. While working in a machine shop you get a spot of grease on your shirt sleeve. In order to remove it, what would you use as a solvent with which to wash the spot?

$$H_2O \qquad C_2H_5OH \qquad C_6H_{14}$$

Explain why.

9. Identify the nature of the following substances as polar or nonpolar.

a. car wax

b. motor oil

c. ammonia

d. carbon dioxide

e. sucrose

f. rubber

g. vaseline

h. rubbing alcohol

i. lipstick

j. gasoline

Experiment 7 Solubility and Miscibility

Name:_____   Section:_____

EXPERIMENT 8

The Synthesis and Properties of Carbon Dioxide, Oxygen, and Hydrogen

I. OBJECTIVES:

A. To synthesize and collect carbon dioxide, oxygen, and hydrogen gases.

B. To determine some of the physical and chemical properties of these gases.

II. DISCUSSION:

The first part of this experiment will involve the production of carbon dioxide. You will produce the carbon dioxide by a double replacement reaction involving sodium hydrogen carbonate and acetic acid:

$$NaHCO_3 \; + \; HC_2H_3O_2 \; \rightarrow \; NaC_2H_3O_2 \; + \; H_2CO_3$$

The carbonic acid that is formed is not a stable compound and undergoes a spontaneous decomposition to carbon dioxide and water:

$$H_2CO_3 \; \rightarrow \; CO_2 \; + \; H_2O$$

The physical properties of the carbon dioxide will be observed after you have collected a sample of the gas. You will also test its flammability and ability to support combustion; two important chemical properties of carbon dioxide.

A third chemical property of carbon dioxide will be observed. You will react the gas with a solution of calcium hydroxide and observe the change that occurs, and relate the change to the following reaction:

$$CO_2 \; + \; Ca(OH)_2 \; \rightarrow \; CaCO_3 \; + \; H_2O$$

Oxygen gas can also be conveniently generated via a simple chemical reaction: the decomposition of dilute hydrogen peroxide solution. When hydrogen peroxide is put on a cut, it bubbles due to the release of oxygen. An enzyme in our blood catalyzes the decomposition of the hydrogen peroxide into water and oxygen. That's why nothing will happen if the hydrogen peroxide is just put on skin that is not cut. A catalyst is a substance that increases the rate at which a reaction occurs without being consumed in the reaction. In our lab experiment, you will use sodium iodide as the catalyst to cause the decomposition of hydrogen peroxide.

$$2H_2O_2 \; \rightarrow \; 2H_2O \; + \; O_2$$

Experiment 8 The Synthesis of CO_2, O_2, and H_2

The physical properties of oxygen gas will be observed and then its chemical property of supporting combustion will be tested.

A convenient way to generate hydrogen gas is to allow a reactive metal, e.g., Zn, Al, Mg, or Fe to react with a strong acid such as hydrochloric acid. Bubbles of hydrogen gas are immediately seen on the surface of the metal. In this experiment you will use Zn and HCl to produce hydrogen.

$$Zn + 2\,HCl \rightarrow ZnCl_2 + H_2$$

The physical properties of the gas will be observed and then its chemical property of combustibility will be tested.

III. MATERIALS:

An 8" test tube equipped with a stopper and rubber tubing, two 6" test tubes with solid stoppers, water trough, wooden splints, burette clamp, saturated $Ca(OH)_2$ solution, 3% H_2O_2 solution, zinc metal, sodium bicarbonate, sodium iodide, acetic acid, 6 M HCl, 10 mL graduated cylinder, scoopula, Bunsen burner, and striker.

IV. PROCEDURE:

A. The Synthesis of Carbon Dioxide:

$$NaHCO_3 + HC_2H_3O_2 \rightarrow NaC_2H_3O_2 + H_2O + CO_2$$

1. Loosely clamp the 8" test tube vertically on a vertical rod of the lattice framework or a ring stand.

2. Fill your trough about half full of water.

3. Fill a 6" test tube full of water, put your thumb over the open end, and immerse the tube, upside down, in the trough, making sure that no air enters the tube. Repeat this procedure for the other 6" test tube.

4. To the 8" test tube add a scoop of sodium bicarbonate **($NaHCO_3$)**.

5. To this reactant add about 5 mL of acetic acid **($HC_2H_3O_2$)** and agitate the tube in order to mix the reactants.

6. Attach the stopper with the rubber tubing to the 8" test tube, wait about 10 seconds to allow the air to be purged out of the test tube and tubing, and then place the end of the tubing under the open end of one of the 6" test tubes. Agitate the 8" test tube periodically to mix the reactants and increase the rate of gas production.

7. Allow the carbon dioxide gas to displace the water in the test tube. When gas collection is complete, stopper the 6" test tube while it is still under the water.

8. Remove the tube and set it aside. Fill the second 6" test tube in the same fashion and then stopper it. Dry the test tubes with a paper towel.

Experiment 8 The Synthesis of CO_2, O_2, and H_2

9. What physical property of carbon dioxide is apparent?

_____

10. Ignite a wooden splint, remove the stopper from one of the 6" test tubes, and plunge the burning splint into the carbon dioxide. What chemical property of carbon dioxide is apparent?

_____

11. To the second 6" test tube add about 5 mL of saturated calcium hydroxide $[Ca(OH)_2]$ stock solution to the tube. Stopper the tube and shake the mixture for about 15 seconds. What do you observe?

_____

12. Remove the rubber tubing and stopper from the 8" reaction tube. Take the tube over to the sink and add about 1 mL of water. What do you observe?

_____

13. Now add about 10 mL of water to the test tube. What happens?

_____

14. Pour the contents of the test tubes down the drain and wash out all of the equipment with soap and water for use in the next part of the experiment.

B. The Synthesis of Oxygen:

$$2 H_2O_2 \rightarrow H_2O + O_2$$

1. Assemble the same apparatus that you used in the carbon dioxide synthesis. Fill a 6" test tube with water as before.

2. To the 8" test tube add about 10 mL of 3% hydrogen peroxide (H_2O_2) stock solution.

3. To the hydrogen peroxide solution in the 8" test tube, add about 1 g of sodium iodide (**NaI**).

4. Immediately attach the stopper the 8" test tube, allow the air to be purged out of the apparatus for about 10 seconds, and then begin to collect the gas in the 6" test tube.

5. When the receiving tube is full of gas, remove the rubber tubing from the bottom of the 6" test tube, and stopper the tube while it's still under the water.

6. Remove the tube from the water and dry it off with a paper towel. What physical property of oxygen is apparent?

_____

7. Turn the test tube right side up.

Experiment 8 The Synthesis of CO_2, O_2, and H_2

8. Ignite a wood splint and let it burn for about 10 seconds. Blow out the flame, immediately remove the rubber stopper from the test tube, and plunge the glowing splint into the oxygen. What chemical property of oxygen is apparent?

_____

9. Wet the wood splint and discard it in splint in the trash container. Pour the contents of the test tube down the drain. Wash the equipment with soap and water in the sink for use in the next part of the experiment.

C. The Synthesis of Hydrogen:

$$Zn + 2\,HCl \rightarrow ZnCl_2 + H_2$$

1. Assemble the same apparatus that you used in the oxygen synthesis. Fill a 6" test tube with water as before. Add about 1 g of zinc (**Zn**) to the 8" test tube.

2. Add about 10 mL of 6 M hydrochloric acid (**HCl**) to the test tube containing the zinc. Hydrogen gas will immediately begin to be generated.

3. Attach the rubber stopper to the 8" test tube, wait about 10 seconds, and then place the end of the tubing under the open end of the 6" test tube.

4. Allow the hydrogen gas to displace the water in the test tube.

5. When all the water has been displaced, remove the rubber tubing, stopper the 6" test tube while it is still under the water, and then remove it from the water while keeping it upside down.

6. Dry the 6" test tube with a paper towel. What property of hydrogen is apparent?

_____

7. Ignite your Bunsen burner.

8. While keeping the 6" test tube inverted, remove the rubber stopper. Ignite a wooden splint with your Bunsen burner and hold the test tube under the burning splint. Then quickly turn the tube right side up so that its **open end is directly beneath the flame of the burning splint. This must be done quickly in order to obtain the proper effect.** What properties of hydrogen are noticed?

Physical property: _____

Chemical property: _____

9. Discard the Zn/HCl reaction mixture in the waste container provided by the instructor, and then rinse out the equipment with water.

Experiment 8 The Synthesis of CO_2, O_2, and H_2

Name:_____   Section:_____

Experiment 8 Problems

1. In this experiment, zinc was reacted with hydrochloric acid to produce hydrogen. Write the balanced equation for the similar reaction that takes place between aluminum and hydrochloric acid.

2. When testing hydrogen gas, why did the tube have to be kept up side down until it was ready to be tested?

3. What are three physical properties of carbon dioxide?

   a.

   b.

   c.

4. What are two chemical properties of carbon dioxide?

   a.

   b.

Experiment 8 The Synthesis of CO_2, O_2, and H_2

5. Consider the following methods of generating carbon dioxide.

$$2\,HCl + Na_2CO_3 \rightarrow 2\,NaCl + H_2O + CO_2$$

$$HCl + NaHCO_3 \rightarrow NaCl + H_2O + CO_2$$

Calculate the amount of CO_2 that can be produced from 5.00 g of sodium carbonate vs. 5.00 g of sodium hydrogen carbonate.

a. 5.00 g of sodium carbonate

b. 5.00 g of sodium hydrogen carbonate

6. Oxygen can be produced by the heating of potassium chlorate, $KClO_3$. How many grams of $KClO_3$ are necessary to produce 25.0 g of oxygen?

$$2\,KClO_3 \rightarrow 2\,KCl + 3\,O_2$$

7. You prepared hydrogen gas from zinc and hydrochloric acid in this experiment. Magnesium reacts with hydrochloric acid in like fashion. Write the balanced equation for this reaction. If you need to prepare 5.00 g of hydrogen, how much magnesium would you have to weigh out? Assume that you have sufficient hydrochloric acid to consume all of the magnesium.

8. If 100.0 g of zinc is mixed with excess hydrochloric acid, how much hydrogen is produced?

9. Water has plenty of oxygen in it. Why doesn't water support combustion?

10. List four physical properties of zinc.

   a.

   b.

   c.

   d.

Experiment 8 The Synthesis of CO_2, O_2, and H_2

Name:_____   Section:_____

EXPERIMENT 9

The Synthesis and Properties of Ammonia

I. OBJECTIVES:

A. To prepare ammonia gas.

B. To determine some of the physical and chemical properties of ammonia.

C. To use red litmus paper to identify the presence of a base.

D. To investigate Graham's Law of Effusion

II. DISCUSSION:

Ammonia gas (NH_3) can be conveniently prepared by heating a mixture of ammonium chloride and calcium hydroxide. A double replacement reaction occurs as follows:

$$2\,NH_4Cl \;+\; Ca(OH)_2 \;\rightarrow\; NH_3 \;+\; H_2O \;+\; CaCl_2$$

If this mixture of gases were collected in a test tube, the water would condense to a liquid and then dissolve some of the ammonia, which is extremely soluble in water. Hence, the water must be removed from this reaction mixture. It can be removed from the ammonia by passing the mixture of gases over calcium oxide. The calcium oxide acts as a desiccant, i.e., it dries the gas mixture by removing the water vapor, without affecting the ammonia, according to the following equation:

$$CaO \;+\; H_2O \;\rightarrow\; Ca(OH)_2$$

The dry ammonia gas can now be tested for its physical and chemical properties.

Litmus is an organic substance obtained from various species of lichens. In acid solutions it is red in color, while in basic solutions it turns blue. You will use paper that has been treated with litmus in its acid form (red) to test a variety of solutions for basic characteristics. Red litmus paper will also be used to test for the presence of ammonia when you are collecting it in test tubes.

Experiment 9 The Synthesis and Properties of Ammonia

III. MATERIALS:

Two 6" test tubes with stoppers, an 8" test tube with stopper and rubber tubing, Bunsen burner, striker, buret clamp, 50 mL and 400 mL beakers, glass wool, red litmus paper, glass rod, scoopula, wood splints, three Q-tips, a piece of glass tubing, ammonium chloride, calcium hydroxide, calcium oxide, concentrated ammonia, concentrated hydrochloric acid, calcium hydroxide solution, sodium hydroxide solution, dilute ammonia solution, sodium chloride solution, and phenolphthalein solution.

IV. PROCEDURE:

A. Testing Solutions with Red Litmus Paper:

To individual pieces of red litmus paper, apply a spot of a solution of calcium hydroxide, sodium hydroxide, ammonia solution, and sodium chloride. What do you observe?

Calcium hydroxide ($Ca(OH)_{2(aq)}$):_____

Sodium hydroxide ($NaOH(aq)$): _____

Ammonia Solution ($NH_3(aq)$): _____

Sodium chloride ($NaCl_{(aq)}$):_____

All compounds that contain the hydroxide ion are bases. What is a common chemical property of bases?

_____

B. The Synthesis of Ammonia:

1. To a 50 mL beaker, add about 2 scoopulafuls of ammonium chloride (NH_4Cl) and 2 scoopulafuls of calcium hydroxide [$Ca(OH)_2$]. With your scoopula, mix the chemicals thoroughly. Transfer the mixture to an 8" test tube and cover the mixture with a piece of glass wool.

2. Clamp the test tube in a horizontal position and carefully place a scoopulaful of calcium oxide (CaO) about half way in the test tube.

3. Attach the rubber stopper with tubing. Hold the end of the rubber tubing about halfway up an inverted 6" test tube and begin vigorously heating the reaction mixture with a Bunsen burner. Ammonia gas will be produced and will start to fill the tube.

4. Since the ammonia is colorless and you cannot see when the tube has been filled, you must chemically determine this. Hold a piece of moist red litmus paper at the bottom of the test tube. When the tube is full of ammonia and it starts to come out of the bottom, the red litmus will turn blue.

5. When the test tube is full of ammonia, remove it and stopper it. Fill another 6" test tube with ammonia gas and stopper it.

6. Turn off the Bunsen burner and allow the reaction tube to cool while you perform the next part of the experiment.

Experiment 9 The Synthesis and Properties of Ammonia

C. Testing the Ammonia:

1. Rinse a 400 mL beaker with distilled water. Fill the beaker about half full of distilled water and add 1 mL of a solution of phenolphthalein. Stir this mixture with a clean glass rod. Take one of the test tubes of ammonia and turn it **up side down**. Remove the stopper and place it in the beaker of water. Allow it to stand in the water while you perform next test. Periodically observe the tube. What do you see occurring?

_____

2. Put a few drops of concentrated hydrochloric acid (**HCl**), on the cotton of a Q-tip. Remove the stopper from one of your ammonia test tubes, and **immediately** insert the Q-tip into the test tube. What do you see occurring?

_____

3. Remove the glass wool and reactants from the tube that was used to generate the ammonia gas. Wash this tube, and the other glassware you used, with soap and water in the sink. Make sure you dispose of the glass wool and the litmus paper in the trash can. Rinse the Q-tip in the sink before disposing of it in the trash can.

D. Graham's Law of Effusion:

1. Obtain a long glass tube and two Q-tips. Wet one Q-tip with a few drops of concentrated ammonia water and another Q-tip with a few drops of concentrated hydrochloric acid. **As simultaneously as possible**, insert the Q-tips into the two ends of the glass tube.

2. Allow the glass tube to sit on your bench top for a minute or two until you notice the formation of a smoke ring. When this ring forms immediately make the measurements in the next step.

3. Using a ruler, measure the distance from each Q-tip to the center of the smoke ring and record these values below.

Distance from ammonia end: _____ cm

Distance from HCl end: _____ cm

3. Rinse the glass tube in the sink with water. Rinse the Q-tips with water and dispose of them in the trash can. Return all the equipment to the appropriate areas.

Experiment 9 The Synthesis and Properties of Ammonia

Name:_____   Section:_____

Experiment 9 Problems

1. List three physical properties of ammonia.

   a.

   b.

   c.

2. List three chemical properties of ammonia.

   a.

   b.

   c.

3. In the previous experiment, you collected hydrogen, oxygen, and carbon dioxide by the downward displacement of water. Why didn't you collect ammonia gas in the same way?

4. Write the reaction between ammonia and hydrogen chloride. Hint: This is a combination reaction which produces a common ionic compound.

5. It has been proposed that ammonia could be used as a fuel in automobiles. It reacts with oxygen in the presence of copper as a catalyst, to produce nitrogen and water. Write a balanced equation that describes this reaction.

6. Why did the water rise in the test tube of ammonia during your first test of the gas?

Experiment 9 The Synthesis and Properties of Ammonia

7. What color change occurred when the water containing the phenolphthalein came in contact with the ammonia gas?

8. You measured the distances of the smoke ring from the Q-tip with ammonia solution and the Q-tip with hydrochloric acid. Calculate the ratio of the two distances by dividing both of them by the smallest of the two and express it in ratio form as X : 1.

9. According to Graham's Law of Diffusion, the ratio of the average velocities, V, of two gases (A and B) at the same temperature is proportional to the inverse of the square roots of their masses, m. Expressed mathematically, this is:

$$\frac{V_A}{V_B} = \frac{\sqrt{m_B}}{\sqrt{m_A}}$$

Calculate the theoretical ratio of velocities of ammonia to hydrogen chloride using this equation, and express your answer as Velocity of Ammonia : Velocity of Hydrogen Chloride is X : 1. For the masses of the two gases, use the formula weights. What relationship do you think exists between the mass and average velocity of the molecules of two gases that are at the same temperature?

10. Using the concept of Graham's Law of Diffusion, arrange the following gases in order of decreasing average molecular velocity (assume that the gases are at the same temperature).

a. hydrogen chloride
b. argon
c hydrogen
d. nitrogen

e. oxygen
f. chlorine
g. ammonia
h. fluorine

_____ > _____ > _____ > _____ > _____ > _____ > _____ > _____

Name:_____   Section:_____

EXPERIMENT 10

Precipitation Reactions and Filtration

I. OBJECTIVES:

A. To understand what is meant by the term "like dissolves like."

B. To predict if a given ionic compound is soluble in water.

C. To mix two aqueous solutions and determine whether or not a precipitate is formed.

D. To be able to calculate a limiting reagent.

E. To be able to calculate percentage yield.

F. To carry out a double replacement reaction and collect the resulting precipitate by filtration.

II. DISCUSSION:

A valuable guide for predicting if a solute will dissolve in a solvent is "like dissolves like." Solutes that are nonpolar will tend to dissolve in solvents that are nonpolar. For example, naphthalene, $C_{10}H_{8\,(s)}$, dissolves readily in hexane, $C_6H_{14\,(l)}$, since both are nonpolar hydrocarbons. Temporary, induced dipoles (London forces) are the interparticle forces of attraction that cause the naphthalene molecules to "stick" to each other. Likewise, the hexane molecules are attracted to each other by London forces. When the solute dissolves in the solvent, the solute molecules must insert themselves between the solvent molecules. This disrupts the London forces. This disruption must be compensated for by interparticle forces of attraction between solute and solvent molecules. Since both compounds are nonpolar, they attract each other by London forces, and a solution results.

Solutes that are polar will tend to dissolve in solvents that are polar. For example, ethyl acetate will dissolve in acetone, since both are polar compounds.

ethyl acetate   acetone

Both of these compounds are liquids whose molecules are attracted to each other by dipole-dipole forces. These same attractions exist between the molecules of acetone and those of ethyl acetate. The result is the solubility of ethyl acetate in acetone.

A form of dipole-dipole attraction that is exceptionally strong is termed **hydrogen bonding**. It occurs among the molecules of compounds that contain H-O, H-N, or H-F bonds. Since these bonds are extremely polar, the hydrogen of one molecule is attracted to the oxygen, nitrogen, or fluorine of another molecule. For example, water molecules are held to each other by many hydrogen bonding attractions:

Solutes that exhibit hydrogen bonding will dissolve in solvents that hydrogen bond, since hydrogen bonding can also occur between solute and solvent molecules. Thus, butyl alcohol $CH_3CH_2CH_2CH_2OH$, will dissolve in ethyl alcohol, CH_3CH_2OH. Additionally, ammonia, NH_3, is extremely soluble in water, due to strong hydrogen bonding forces between these two substances.

Substances that hydrogen bond are also able to form solutions with nonhydrogen bonding polar substances that contain oxygen and nitrogen. This is due to the attraction of the hydrogen of the hydrogen bonding molecule to the oxygen or nitrogen of the polar molecule. Thus, acetone is soluble in water due to hydrogen bonding.

The third type of compound (after nonpolar and polar) is ionic. As you might expect, ionic compounds dissolve in ionic compounds. Since ionic compounds are solids at room temperature, these types of solutions occur only at elevated temperatures where at least one of the compounds is a liquid. If an ionic solute is mixed with a nonpolar or polar solvent, it will normally not dissolve. An important exception to this guideline is water. Due to its extremely high polarity (hydrogen bonding), water is able to dissolve **some** ionic compounds. This is due to its ability to form ion-dipole attractions between some ions and water. For example, sodium chloride is soluble in water due to these attractions:

Note that the attraction is always between a cation and the oxygen, and between an anion and the hydrogen.

Experiment 10 Precipitation Reactions and Filtration

A simple method for determining if an ionic compound is water soluble, is to look for the presence of particular ions. **Table 1** summarizes these guidelines.

Table 1

Compound is soluble if it contains the following ion	Exceptions
NH_4^+	None that are important
Group IA metal cations	None that are important
NO_3^-	None that are important
$C_2H_3O_2^-$	None that are important
Cl^-	$AgCl$, $PbCl_2$, Hg_2Cl_2
SO_4^{2-}	$PbSO_4$, $BaSO_4$, $CaSO_4$, Ag_2SO_4

Thus, the following compounds would be water soluble due to the presence of the above ions: $Al(NO_3)_3$, $(NH_4)_2CO_3$, KBr, $Cu(C_2H_3O_2)_2$, $CaCl_2$, $MgSO_4$, and Na_3PO_4. Due to the lack of the above ions, the following compounds are insoluble in water: $PbCO_3$, $Mg_3(PO_4)_2$, $BaSO_4$, $AgCl$, CuS, and $Al(OH)_3$.

When an aqueous solution of an ionic compound is poured into an aqueous solution of a second ionic compound, two different results may be obtained. A reaction (generally termed a **double replacement reaction**) may occur to produce a new substance, or a simple mixing of ions may result. The outcome depends on the solubility of the resulting potential products. Consider mixing a solution of potassium chloride with a solution of sodium nitrate.

$$KCl + NaNO_3 \rightarrow KNO_3 + NaCl$$

Does this really represent a chemical reaction? The answer can be more clearly seen if we write this as an **ionic equation**. Since all of the reactants and products are water soluble, their solutions consist of individual ions dispersed in water, i.e.,

$$K^+(aq) + Cl^-(aq) + Na^+(aq) + NO_3^-(aq) \rightarrow K^+(aq) + NO_3^-(aq) + Na^+(aq) + Cl^+(aq)$$

Notice that there is no difference between the left and right side of this equation, except the order in which the ions are written. Hence, when these solutions are combined, only the physical process of mixing occurs; not a chemical reaction.

When does a chemical reaction occur between two ionic compounds? Consider mixing a solution of sodium chloride with a solution of silver nitrate, i.e.,

$$NaCl + AgNO_3 \rightarrow NaNO_3 + AgCl$$

Written as an ionic equation, we have:

$$Na^+(aq) + Cl^-(aq) + Ag^+(aq) + NO_3^-(aq) \rightarrow Na^+(aq) + NO_3^-(aq) + AgCl(s)$$

A chemical change has occurred in this case, due to the formation of an insoluble product, the silver chloride.

A generalization that we can make concerning the mixing of ionic solutions, is that a reaction will occur when an insoluble product is formed. Which of the following equations represents a chemical change, and not just a physical mixing of solutions?

$$CuSO_4 + 2NaC_2H_3O_2 \rightarrow Cu(C_2H_3O_2)_2 + Na_2SO_4$$

$$2(NH_4)_3PO_4 + 3Zn(NO_3)_2 \rightarrow 6NH_4NO_3 + Zn_3(PO_4)_2$$

If you said the second one, you are absolutely correct. It represents a chemical reaction since the zinc phosphate is insoluble, and precipitates as it is formed. In the first equation, no reaction occurs since both of the formulas that are written represent compounds that are water soluble. In this case, only a mixing of ions has resulted.

In this experiment, you will mix a variety of aqueous solutions of ionic compounds and determine if a reaction occurs (as evidenced by the formation of a precipitate). In another part of the experiment you will carry out the following reaction:

$$CaCl_2(aq) + Na_2CO_3(aq) \rightarrow 2NaCl(aq) + CaCO_3(s)$$

A known mass of calcium chloride and sodium carbonate will be dissolved in water, and the resulting solutions combined. After the precipitation has been formed, it will be collected by filtration, dried, and weighed. The yield that you obtain will be compared to the **theoretical yield**.

The theoretical yield is calculated by comparing the amount of calcium carbonate that could be obtained from the calcium chloride (assuming excess sodium carbonate) with the amount that could be obtained from the sodium carbonate (assuming excess calcium chloride). This is performed in the following fashion:

$$\text{grams CaCl}_2 * \frac{1 \text{ mole CaCl}_2}{\text{Mol Wt CaCl}_2} * \frac{1 \text{ mole CaCO}_3}{1 \text{ mole CaCl}_2} * \frac{\text{Mol Wt CaCO}_3}{1 \text{ mole CaCO}_3} = \text{grams CaCO}_3$$

$$\text{grams Na}_2\text{CO}_3 * \frac{1 \text{ mole Na}_2\text{CO}_3}{\text{Mol Wt Na}_2\text{CO}_3} * \frac{1 \text{ mole CaCO}_3}{1 \text{ mole Na}_2\text{CO}_3} * \frac{\text{Mol Wt CaCO}_3}{1 \text{ mole CaCO}_3} = \text{grams CaCO}_3$$

The theoretical yield of calcium carbonate is the smallest value from the preceding two calculations. The reagent that produces this smallest yield is termed the **limiting reagent**.

The **percentage yield** of a reaction is the ratio of the actual yield that you obtain, to the theoretical yield, i.e.,

$$\% \text{ Yield} = \frac{\text{Actual Yield}}{\text{Theoretical Yield}} * 100\%$$

By determining the percentage yield, you will be able to gauge how effectively you were able to transfer and collect the calcium carbonate product. With careful work, you should be able to obtain a relatively high percentage yield.

III. MATERIALS:

Two 250 mL beakers, calcium chloride dihydrate, sodium carbonate dodecahydrate, balance, 100 mL graduated cylinder, 2 glass rods, large watch glass, squeeze bottle of distilled water, thermometer, 12.5 cm filter paper, glass funnel, wire gauze, iron ring, ring stand, Bunsen burner, striker, 250 mL Erlenmeyer flask, oven. Spot plate, dropping bottles with solutions of sodium chloride, sodium sulfate, sodium carbonate, sodium phosphate, potassium nitrate, calcium nitrate, barium nitrate, ammonium nitrate, silver nitrate.

Experiment 10 Precipitation Reactions and Filtration

IV. PROCEDURE:

A. The Precipitation and Isolation of Calcium Carbonate:

1. Tare a 250 mL beaker (Beaker 1) and record its mass in **Table 2**. To the beaker, add about 5 g of calcium chloride dihydrate **($CaCl_2 \cdot H_2O$)**. Weigh the beaker again to get the total mass, and record this in the Table. By difference, determine the mass of the calcium chloride dihydrate. Calculate the mass of calcium chloride in the calcium chloride dihydrate. Record this in the Table.

2. Tare a second 250 mL beaker (Beaker 2) and record its mass in the Table. Add about 13 g of sodium carbonate dodecahydrate **($Na_2CO_3 \cdot 12\ H_2O$)** to this beaker, weigh it again, record its mass, and by difference, determine the mass of the sodium carbonate dodecahydrate. Record these data in the Table. Calculate the mass of the sodium carbonate in the sodium carbonate dodecahydrate. Record this in the Table.

3. To each beaker, add about 50 mL of distilled water. Stir each solution with a glass rod until the solids have completely dissolved.

4. Pour the solution of sodium carbonate into the beaker containing calcium chloride solution. Use about 5 mL of distilled water to rinse the empty beaker, and pour this into the beaker containing the reactants.

5. When the solutions are mixed, a thick, somewhat gelatinous precipitate forms. The crystals of this precipitate are rather small, and may clog the filter paper when you attempt to isolate the product by filtration. The size of the crystals can be increased by a process called **digestion**. Place the beaker on a piece of wire gauze on an iron ring attached to a ring stand. Use a Bunsen burner to heat the mixture to about 95°C. After the mixture reaches this temperature, turn off the burner and allow the mixture to cool to room temperature. If you remove the beaker from the wire gauze and place it on the bench top, it will cool more quickly.

6. While the mixture is cooling, prepare your filtration apparatus. Tare a piece of 12.5 cm filter paper and record its mass in the Table. Fold the filter paper in half and then into quarters, as shown in **Figure 1**.

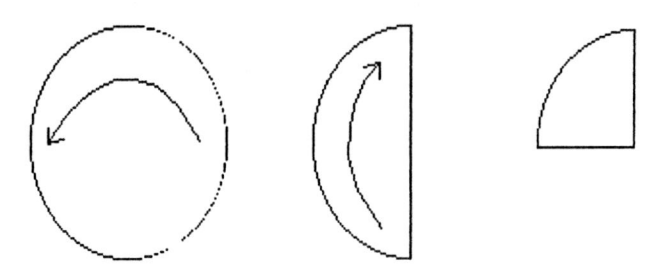

Figure 1

Open the paper so that a cone is formed with one piece of paper on one half and three pieces of paper on the other half. Place the filter paper in a glass funnel, place the funnel in an iron ring clamped above a 250 mL Erlenmeyer flask, and wet the paper with distilled water. It should now stick to the sides of the funnel, as shown in **Figure 2**.

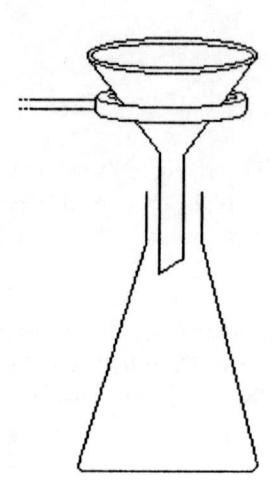

Figure 2

7. After the reaction mixture has cooled to room temperature, decant the supernatant liquid into the filter paper by allowing it to run down a glass rod, as shown in the following **Figure 3**.

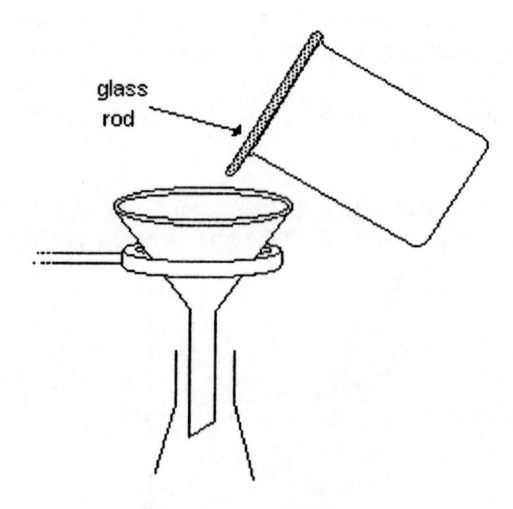

Figure 3

8. Using a wash bottle of distilled water, rinse all the precipitate out of the beaker into the filter paper. After the solution has stopped draining from the beaker, rinse the precipitate with two 10 mL portions of distilled water. After the water has stopped draining, transfer the filter paper to a watch glass, and place it in an oven to dry. While the product is drying, perform **Part C** of this experiment.

9. When the paper and precipitate are dry, reweigh them, and record this value in the Table. By difference determine the mass of the calcium carbonate. Record this value.

Experiment 10 Precipitation Reactions and Filtration

Table 2

Mass of Empty Beaker 1	
Mass of Beaker 1 and $CaCl_2 \cdot 2\,H_2O$	
Mass of $CaCl_2 \cdot 2\,H_2O$	
Mass of $CaCl_2$	
Mass of Empty Beaker 2	
Mass of Beaker 2 and $Na_2CO_3 \cdot 12\,H_2O$	
Mass of $Na_2CO_3 \cdot 12\,H_2O$	
Mass of Na_2CO_3	
Mass of Filter Paper	
Mass of Filter Paper and Precipitate	
Mass of Precipitate	

B. The Calculation of Theoretical and Percentage Yields:

1. Calculate the theoretical yield of calcium carbonate that could be obtained from your mass of calcium chloride. Record the value in **Table 3**. Show your work in the space below.

2. Calculate the theoretical yield of calcium carbonate that could be obtained from your mass of sodium carbonate. Record this value in the Table. Show your work in the space below.

3. Which of these two values is the correct theoretical yield of your reaction? _____ g

What is the limiting reagent in this reaction? _____

4. Calculate the percentage yield of your product and record it in the Table.

Table 3

Theoretical Yield of $CaCO_3$ from $CaCl_2$	
Theoretical Yield of $CaCO_3$ from Na_2CO_3	
Percentage Yield of $CaCO_3$	

Experiment 10 Precipitation Reactions and Filtration

C. Double Replacement Precipitation Reactions:

1. Obtain a plastic spot tray and a set of ionic test solutions:

    a. sodium chloride **(NaCl)**

    b. sodium sulfate **(Na$_2$SO$_4$)**

    c. sodium carbonate **(Na$_2$CO$_3$)**

    d. sodium phosphate **(Na$_3$PO$_4$)**

    e. potassium nitrate **(KNO$_3$)**

    f. calcium nitrate **[Ca(NO$_3$)$_2$]**

    g. barium nitrate **[Ba(NO$_3$)$_2$]**

    h. ammonium nitrate **(NH$_4$NO$_3$)**

    i. silver nitrate **(AgNO$_3$)**

2. To the upper-left-hand well of the spot tray, add two drops of sodium chloride solution and two drops of potassium nitrate solution. Continue adding two drops of each of the reactants to the other wells for the combinations shown in **Table 4**. In this Table, record whether you obtain only a solution **(S)** or an insoluble product **(I)**. If it is difficult to see a precipitate in a well, try looking from beneath the tray. **Do not place the tray directly above your head.**

Table 4

	NaCl	Na$_2$SO$_4$	Na$_2$CO$_3$	Na$_3$PO$_4$
KNO$_3$				
Ca(NO$_3$)$_2$				
Al(NO$_3$)$_3$				
NH$_4$NO$_3$				
AgNO$_3$				

3. For the wells in which a precipitate formed, determine the formula of the product with the help of the solubility rules, and record it in **Table 5**.

Table 5

	NaCl	Na_2SO_4	Na_2CO_3	Na_3PO_4
KNO_3				
$Ca(NO_3)_2$				
$Al(NO_3)_3$				
NH_4NO_3				
$AgNO_3$				

4. When you are finished with both parts of this experiment, clean all of your glassware thoroughly with soap and water and return all equipment to the designated area. Be especially careful to clean the wells of the spot plate in which a precipitate was formed. Use a test tube brush to remove any traces of precipitate.

Experiment 10 Precipitation Reactions and Filtration

Name:_____   Section:_____

Experiment 10 Problems

1. What three types of attractive forces are involved in solution formation?

   a.

   b.

   c.

2. Explain why cations in aqueous solutions are attracted to the oxygens of water molecules and why anions are attracted to the hydrogens of water molecules.

3. Which of the following pairs of compounds would form a solution?

   a. $C_{20}H_{42}$ and H_2O e. CO_2 and H_2O

   b. C_5H_{12} and C_7H_{16} f. CO_2 and C_7H_{16}

   c. HCl and H_2O g. CCl_4 and H_2O

   d. CH_3OH and H_2O h. CH_3NH_2 and H_2O

4. Which of the following ionic compounds are water soluble?

   a. $CsCl$ g. $Mn_3(PO_4)_2$ m. $NaClO_4$

   b. K_2CO_3 h. $(NH_4)_2Cr_2O_7$ n. $FeCl_3$

   c. $MgSO_4$ i. $U(NO_3)_3$ o. $KC_2H_3O_2$

   d. $AgCl$ j. $(NH_4)_3PO_4$ p. $NiSO_4$

   e. $Ni(NO_3)_3$ k. Cr_2S_3 q. $Al(OH)_3$

   f. CuS l. $CuCO_3$ r. $NaOH$

Experiment 10 Precipitation Reactions and Filtration

5. Which of the following equations represents actual chemical reactions?

   a. $2 \text{ KCl} + \text{Pb(NO}_3)_2 \rightarrow \text{PbCl}_2 + 2 \text{ KNO}_3$

   b. $3 \text{ KC}_2\text{H}_3\text{O}_2 + \text{Ni(NO}_3)_3 \rightarrow 3 \text{ KNO}_3 + \text{Ni(C}_2\text{H}_3\text{O}_2)_3$

   c. $\text{Na}_2\text{CO}_3 + \text{Cu(NO}_3)_2 \rightarrow 2 \text{ NaNO}_3 + \text{CuCO}_3$

   d. $3 \text{ MgCl}_2 + 2 \text{ Na}_3\text{PO}_4 \rightarrow \text{Mg}_3(\text{PO}_4)_2 + 6 \text{ NaCl}$

   e. $2 \text{ Fe(NO}_3)_3 + 3 \text{ CaSO}_4 \rightarrow \text{Fe}_2(\text{SO}_4)_3 + 3 \text{ Ca(NO}_3)_2$

   f. $2 \text{ AgNO}_3 + \text{Na}_2\text{SO}_4 \rightarrow \text{Ag}_2\text{SO}_4 + 2 \text{ NaNO}_3$

6. Write ionic equations for the following:

   a. $\text{Na}_3\text{PO}_4 + 3 \text{ AgNO}_3 \rightarrow 3 \text{ NaNO}_3 + \text{Ag}_3\text{PO}_4$

   b. $(\text{NH}_4)_2\text{SO}_4 + \text{Ba(C}_2\text{H}_3\text{O}_2)_2 \rightarrow 2 \text{ NH}_4\text{C}_2\text{H}_3\text{O}_2 + \text{BaSO}_4$

7. What does it mean to digest a precipitate?

8. What are some possible reasons for the actual yield of a reaction being lower than the theoretical yield?

   a.

   b.

   c.

Experiment 10 Precipitation Reactions and Filtration

9. Acid rain is produced, in part, by the reaction of nitrogen dioxide with rain water, according to the following equation:

$$3\ NO_2 + H_2O \rightarrow 2\ HNO_3 + NO$$

What weight of nitric acid can be produced from 18.5 g of nitrogen dioxide?

10. Chlorine gas can be produced from the reaction of manganese(IV) oxide with hydrochloric acid as shown in the following equation:

$$MnO_2 + 4\ HCl \rightarrow MnCl_2 + 2\ H_2O + Cl_2$$

How many grams of chlorine can be produced from 76.0 g of MnO_2 (assume sufficient HCl)?

11. If 18.2 g of a product could theoretically be produced in a reaction, but only 14.6 g of it are actually obtained, what is the percentage yield of the reaction?

12. Given the reaction:

$$2\ Al + 3\ H_2SO_4 \rightarrow Al_2(SO_4)_3 + 3\ H_2$$

If an excess of sulfuric acid is combined with 10.0g of aluminum, and 0.150 g of hydrogen is obtained, what is the percentage yield of the hydrogen?

Name:_____   Section:_____

EXPERIMENT 11

The Qualitative Analysis of an Ionic Compound

I. OBJECTIVES:

A. To observe the flame emission color from excited metal ions.

B. To identify what cation is present in an unknown on the basis of a flame test.

C. To observe some reactions of common anions.

D. To use blue litmus to determine the acidity of a gas.

E. To identify what anion is present in an unknown substance.

II. DISCUSSION:

A. Qualitative Analysis:

An important aspect of chemistry is the identification of the composition of an unknown substance. The area of chemistry that is concerned with the identification of substances is referred to as **qualitative analysis**. Substances are identified by the use of spectral characteristics, by reactions that are characteristic of the substance, or by both. In this experiment, you will be given an unknown ionic substance and asked to determine what cation and anion are present in the compound. The cation will be determined by its production of a characteristic color when it is heated. The anion that accompanies the cation will be determined by using a series of chemical reactions.

B. Flame Tests:

The cation in your unknown, e.g., sodium, potassium, copper, calcium, etc., can easily be determined by using a **flame test**. In this procedure, a sample of the compound, usually dissolved in distilled water, is heated in the flame of a Bunsen burner. Upon heating, a color flares up from the sample that is characteristic of the particular cation being tested.

The electrons that surround the nucleus of an atom occupy specific energy levels. Those energy levels closer to the nucleus are of lower energy, in general, than those farther away from the nucleus. The most stable arrangement of the electrons in an atom is referred to as the atom's **ground state**. When an atom absorbs energy, it absorbs energy of certain amounts, i.e., quantum amounts, so that an electron is promoted to a higher energy level. Occupation of such an energy level by an electron is more difficult to maintain, and the atom is said to be in an unstable **excited state**. When the electron returns to the ground state from the excited state, it releases the same amount of energy previously absorbed, but in the form of X-rays, ultraviolet, infrared, and visible light. These are some common examples of electromagnetic radiation. The electromagnetic spectrum is show in the following figure.

The Electromagnetic Spectrum

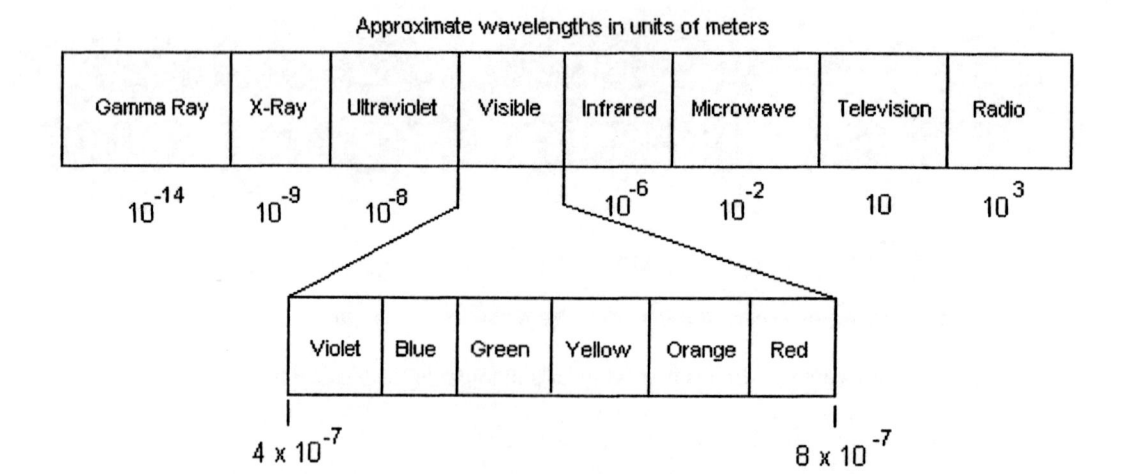

Approximate wavelengths in units of meters

Gamma Ray	X-Ray	Ultraviolet	Visible	Infrared	Microwave	Television	Radio
10^{-14}	10^{-9}	10^{-8}		10^{-6}	10^{-2}	10	10^3

Violet	Blue	Green	Yellow	Orange	Red

4×10^{-7} 8×10^{-7}

The electron energy levels (or states) of the atoms of one element are different from those in the atoms of other elements. For example, the 19 electrons in a potassium atom have a different set of energy states than the 56 electrons in barium. In this experiment you will use the heat of a flame to cause the excitation of electrons in different types of atoms. When the electrons return to lower energy levels, a color (specific for each element and therefore, a means of identification) will be released. You will use this as the basis of identifying the cation that is present in your unknown.

C. Chemical Tests for Anions:

A chemical reaction that is characteristic of a substance is referred to as a **chemical test**. For example, you tested for oxygen with a glowing splint. If the splint bursts into flame, oxygen is present. Chloride ions can be tested for by the use of concentrated sulfuric acid. If chloride ions are present, hydrogen chloride gas bubbles out of the reaction mixture. However, since other substances may yield a gas under these conditions, the presence of the hydrogen chloride has to be confirmed by doing a second chemical test. If the second test is also positive, the presence of the chloride ion is almost certain. A convenient way of testing a gas is by the use of litmus paper. As you saw in the ammonia experiment, ammonia gas causes red litmus to turn blue. This is due to the ability of the ammonia to react with the water in the litmus paper and produce a small amount of hydroxide ions. It is these ions that cause the color change of the red litmus. In this experiment, hydrogen chloride gas will cause blue litmus to turn red. Generally, substances whose formula begins with hydrogen, e.g. HCl, produce an acid when dissolved in water. It is this acid that causes the color change from blue to red. Other gases may not have any effect on the colors of litmus.

In this experiment, you will perform chemical tests using many substances which are found in everyday life, such as table salt, Epsom salt, battery acid, and baking powder. After observing the results of reactions with these substances, you will test your unknown. By performing similar tests with the unknown, you will be able to determine whether it contains carbonate, chloride, sulfate, or iodide ions.

Experiment 11 The Qualitative Analysis of an Ionic Compound

III. MATERIALS:

A test tube rack, ten 4" test tubes, blue litmus paper, Bunsen burner, striker, a Nichrome wire in a cork, a solid unknown, sodium carbonate, concentrated sulfuric acid, Epsom salt, potassium iodide, silver nitrate solution, solutions containing sodium chloride, potassium chloride, copper chloride, barium chloride, strontium chloride, lithium chloride.

IV. PROCEDURE:

A. Flame Tests of Known Cations:

Work in pairs in this experiment. Ignite your Bunsen burner and adjust it to give a flame with as little color as possible. Obtain a bottle containing one of the metal ions listed below. Place the wire, with some of the solution adhering to it, in the flame and record the color of the flame in **Table 1**. The color to record is the one that appears on the initial "flare up." If the wire is held in the flame after this initial flame, a color will appear due to the wire itself. If the color which you observe is not clear or apparent, do the test again. Several metals give very similar colors. Be very careful in noting the difference in the colors of these metals. Repeat the test for the five other cation samples.

Table 1

Metal Cation	Flame Test Observations
Sodium	
Potassium	
Copper	
Barium	
Strontium	
Lithium	

B. Flame Test of an Unknown:

Obtain an unknown from the instructor. Using the nichrome wire that is held in a rubber stopper, carry out a flame test on this solution just as you did with the known solutions. If it is difficult to tell between two metals, you may want to go back and redo the two cations, looking for more subtle differences.

Letter of the unknown: _____

Name of the metal cation in your unknown: _____

C. Testing for the Presence of Carbonate Ion in Na_2CO_3:

1. Testing with Sulfuric Acid:

$$Na_2CO_3 + H_2SO_4 \rightarrow Na_2SO_4 + H_2O + CO_2$$

Experiment 11 The Qualitative Analysis of an Ionic Compound

Place a small pea-sized amount of sodium carbonate **(Na₂CO₃)** in a small, dry test tube. Have a piece of blue litmus at hand for the next test. Add half a dropperful of concentrated sulfuric acid **(H₂SO₄)** and record your observations in **Table 2**. **Caution: Sulfuric acid is extremely caustic. Avoid contact with your skin or clothes. If you do get sulfuric acid on yourself, wash it off immediately and then see the instructor.**

2. Testing with Blue Litmus Paper:

As the gas is being generated in the above test, hold a piece of blue litmus paper over the mouth of the test tube. **Make sure that the litmus paper does not touch the test tube.** Record your observations.

3. Discard the contents of the test tube down the sink and wash out the tube with soap and water.

D. Testing for the Presence of Chloride Ion in NaCl:

1. Testing with Sulfuric Acid:

$$2\,NaCl \;+\; H_2SO_4 \;\rightarrow\; Na_2SO_4 \;+\; 2\,HCl$$

Place a small pea-sized amount of sodium chloride **(NaCl)** in a small, dry test tube. Have a piece of blue litmus paper at hand for the next test. Add a dropperful of concentrated sulfuric acid **(H₂SO₄)**. **Very carefully** note the color and odor of the escaping gas. **Be sure not to place your nose over the mouth of the test tube.** Record your observations.

2. Testing with Blue Litmus Paper:

Hold a piece of blue litmus paper over the mouth of the test tube in the above test and record your observations. **Make sure that the litmus paper does not touch the test tube.**

3. Discard the contents of the test tube down the sink and wash out the tube with soap and water.

E. Testing for the Presence of Sulfate Ion in MgSO₄:

1. Testing with Sulfuric Acid:

Place a small quantity of Epsom salt **(MgSO₄)** in a small, dry test tube and add a dropperful of concentrated sulfuric acid **(H₂SO₄)**. Record your observations.

2. Testing with Barium Chloride:

$$MgSO_4 \;+\; BaCl_2 \;\rightarrow\; BaSO_4 \;+\; MgCl_2$$

Place a small amount of Epsom salt **(MgSO₄)** in a small test tube and dissolve it in about 1 mL of **distilled water**. Make sure that all of the Epsom salt has dissolved. If necessary, add another milliliter of water. Once it is dissolved, add 2 drops of a barium chloride **(BaCl₂)** stock solution. Record your observations.

3. Discard the contents of the test tubes in the container provided by the instructor and wash the tubes with soap and water.

Experiment 11 The Qualitative Analysis of an Ionic Compound

F. Testing for the Presence of Iodide Ion in KI:

1. Testing with Sulfuric Acid:

$$2\,KI + 2\,H_2SO_4 \rightarrow K_2SO_4 + SO_2 + 2\,H_2O + I_2$$

Place a small amount of potassium iodide **(KI)** in a small, dry test tube and add a dropperful of concentrated sulfuric acid **(H₂SO₄)**. Record your observations below.

2. Discard the contents of the test tube down the sink and wash out the tube with soap and water.

3. Testing with Silver Nitrate:

$$KI + AgNO_3 \rightarrow KNO_3 + AgI$$

Place a small amount of potassium iodide **(KI)** in a small test tube, add about 1 mL of water, and shake the tube to dissolve the solid. Add 2 drops of silver nitrate **(AgNO₃)** stock solution and record your observations.

4. Discard the contents of the test tube in the waste container provided by the instructor and wash out the tube with soap and water.

Table 2

Anion	Observations	
	Test with H$_2$SO$_4$	Confirmatory Test
Carbonate		Blue Litmus:
Chloride		Blue Litmus:
Sulfate		Barium Chloride:
Iodide		Silver Nitrate:

G. Testing the Unknown:

1. You will now perform some chemical tests on your unknown to determine if it contains carbonate, chloride, sulfate, or iodide ions. Since all of the knowns that you tested were reacted with sulfuric acid, that will be your first test.

2. Testing with Sulfuric Acid:

Place a small amount of your unknown in a small, dry test tube. Add a dropperful of concentrated sulfuric acid **(H₂SO₄)**. Record your observation in **Table 3**.

Experiment 11   The Qualitative Analysis of an Ionic Compound

3. Confirmatory Test:

Compare this observation to the earlier ones that you made with the knowns. It should compare favorably to one of them. To make sure you that you have correctly identified the unknown, run a **confirmatory test**, i.e., one of the second tests that you performed on your knowns. For example, if your sulfuric acid test indicated that you have chloride ion present, use the blue litmus test to confirm the presence of HCl gas when the unknown is treated with sulfuric acid. Record the confirmatory test that you chose, and your observations. From your results, determine what anion was present in your unknown, and record its name in the **Table**. Discard the contents of the test tube in the waste container provided by the instructor and wash out the tube with soap and water. Return all chemicals and equipment to the appropriate areas of the lab.

Table 3

Test	Observation
H_2SO_4	
What confirmatory test did you use? _____	
Name of the anion in your unknown:	

H. Identity of the Unknown:

Based on your tests for the anion and cation, what is the formula of your unknown? _____

Experiment 11 The Qualitative Analysis of an Ionic Compound

Name:_____   Section:_____

Experiment 11 Problems

1. A substance turns blue litmus red. Is the substance an acid or base?

2. What kind of color change (if any) would you see when the following substances were tested with litmus paper?

	With red litmus	With blue litmus
a. $NH_3(aq)$	_____	_____
b. $H_2C_2O_4$	_____	_____
c. NaOH	_____	_____
d. HBr	_____	_____
e. $Mg(OH)_2$	_____	_____
f. $HC_2H_3O_2$	_____	_____

3. Write a balanced equation for the reaction between nickel(II) chloride and sulfuric acid.

4. What results would you observe if hexane (C_6H_{14}) was tested with red litmus paper and with blue litmus paper. Explain your observations.

5. A solution of an unknown ionic compound was treated with concentrated hydrochloric acid. A gas was produced which did not affect blue litmus paper. When the solution was subjected to a flame test a bright orange flame was observed. Write the formula of the ionic compound.

6. A solution of an unknown ionic compound was treated with concentrated sulfuric acid with no observable change. When treated with a barium chloride solution, a precipitate was observed. A flame test of this aqueous solution of the ionic compound produced a light violet color. Write the formula of the ionic compound.

Experiment 11 The Qualitative Analysis of an Ionic Compound

7. In this experiment you saw that you could obtain hydrogen chloride gas by reacting sodium chloride and sulfuric acid. If you needed 50.0 g of hydrogen chloride, how much salt would you need to react with sulfuric acid?

8. What are the seven different types of electromagnetic radiation?

a. _____   e. _____

b. _____   f. _____

c. _____   g. _____

d. _____

9. In what energy level are the valence electrons (the outer energy level electrons) of the atoms of the following elements?

   a. Ba   b. Cu   c. Sr   d. Li

10. a. Write the electron configuration for potassium in terms of energy levels.

b. Which electron gets promoted to a higher energy state when an atom of potassium absorbs energy?

Name:_____   Section:_____

EXPERIMENT 12

The Determination of the Molar Weight of Butane

I. OBJECTIVES:

A. To simultaneously determine the mass, volume, pressure, and temperature of a sample of butane gas.

B. By using the above data, the molar weight of butane gas will be calculated.

II. DISCUSSION:

The Ideal Gas Law states that the product of the pressure and volume of a gas is equal to the product of the number of moles of the gas, its Kelvin temperature, and a constant (the Universal Gas Constant). In an equation form, this is:

$$PV = nRT$$

where R is the Universal Gas Constant. Its values are 0.0821 L-atm/mole-K or 62.4 L-mmHg/mole-K. The choice of which value to use depends upon the units of pressure being employed in the experiment.

The number of moles of a substance can be determined by dividing the number of grams of the substance by its molar weight. Again, in an equation form, this is:

$$n = \frac{g}{Mol.\,Wt.}$$

Thus, a modified version of the Ideal Gas Law can be written as:

$$PV = \frac{gRT}{Mol.\,Wt.}$$

If we rearrange this equation for the molar weight, we have:

$$Mol.\,Wt. = \frac{gRT}{PV}$$

Hence, if we simultaneously know the conditions of mass, temperature, pressure, and volume for a sample of gas, we may calculate its molar weight.

Experiment 12   The Determination of the Molar Weight of Butane

In this experiment, you will determine the molar weight of the gas butane, commonly found in inexpensive pocket lighters. You will release butane from the lighter and allow it to displace water in a flask. By measuring the amount of displaced water (which is therefore, the amount of gas released from the butane lighter), the temperature and pressure of the gas, and determining the weight of the gas that was collected, you can calculate the molar weight of the butane gas.

III. MATERIALS:

A 125 mL Erlenmeyer flask, glass plate, trough, butane lighter, thermometer, 100 mL graduated cylinder, and marking pen.

IV. PROCEDURE:

A. Collecting the Butane:

1. Work in pairs in this experiment.

2. Fill a 125 mL Erlenmeyer flask completely full of water. Cover the mouth of the flask with a glass plate, making sure that no air gets trapped in the flask.

3. Fill a trough about 3/4 full of water. While holding the glass plate in place with your finger, carefully turn the flask upside down and place it in the trough. Make sure that no air bubbles enter the flask. Remove the glass plate.

4. Obtain a lighter and wipe it clean and dry it with a paper towel. Blow out any water that may be trapped in the striker mechanism from a previous experiment. If the lighter is adjustable, make sure the flame adjustment is set to give the largest flame.

5. Place the clean, dry lighter on a balance and determine its weight to the nearest hundredth of a gram. Record this value in **Table 1**.

6. Carefully hold the butane lighter in the water underneath the Erlenmeyer flask. Make sure that the gas opening of the lighter is beneath the mouth of the flask. Press the release lever, being sure that the gas bubbles enter the flask. If bubbles escape, the experiment must be started from the beginning.

7. Continue to hold down the lever until you have collected about 125 mL of gas.

8. Remove the butane lighter.

9. Carefully raise or lower the gas-filled flask until the level of the water inside the flask is at the same point as the water level outside the flask. This is to make sure that the pressure inside the flask is the same as the atmospheric pressure outside the flask. Using a marking pen, draw a line on the flask to record the water level.

10. Remove the flask from the trough, turn it right-side-up, and fill it with water to the wax mark.

11. In order to accurately determine the volume of the water in the flask, pour the water into the 100 mL graduated cylinder (in several portions). Record the total water volume in **Table 1**.

12. Thoroughly dry the lighter with a paper towel, weigh it, and record its mass in **Table 1**.

13. Measure the temperature of the water and record its value in **Table 1**.

Experiment 12 The Determination of the Molar Weight of Butane

14. Record the barometric pressure in **Table 1**. Your instructor will provide you with the current pressure reading, or you might be asked to make a pressure reading from the lab barometer.

Table 1

Initial Mass of the Lighter	g
Final Mass of the Lighter	g
Mass of the Butane Collected	g
Volume of the Butane Collected	mL
Temperature of the Water	°C
Atmospheric Pressure	mmHg

B. Calculation of the Molar Weight of Butane:

1. You must first determine the partial pressure of the butane gas. The reason for this is that the gas in the flask is not all from the butane lighter. There is also water vapor present. The total pressure of the gas in the flask is therefore the pressure of the butane plus the pressure of the water vapor (Dalton's Law of Partial Pressure). In other words:

$$P_{total} = P_{butane} + P_{water}$$

Table 2 gives the vapor pressures of water at different temperatures. Select the vapor pressure that corresponds to the temperature which you recorded in **Table 1**. Subtract this value from the total gas pressure (the atmospheric pressure) in order to obtain the pressure of the butane.

$$P_{butane} = P_{total} - P_{water}$$

$$= \underline{\hspace{5cm}} \text{ mmHg}$$

Experiment 12   The Determination of the Molar Weight of Butane

Table 2

Vapor Pressure of Water at Various Temperatures			
Temperature, °C	Pressure, mmHg	Temperature, °C	Pressure, mmHg
15	12.8	23	21.0
16	13.6	24	22.4
17	14.5	25	23.7
18	15.5	26	25.2
19	16.5	27	26.7
20	17.5	28	28.3
21	18.6	29	30.0
22	19.8	30	31.8

2. Now having determined the actual pressure of the butane gas in the flask, and knowing the volume, mass, and temperature of the gas (from **Table 1**), calculate the molar weight of butane gas. Don't forget to change the volume of the butane from mL to L and the temperature from degrees Celsius to Kelvin units. Show all your work in this calculation.

Molar Weight of Butane: _____ g/mole

Experiment 12   The Determination of the Molar Weight of Butane

Name:_____   Section:_____

Experiment 12 Problems

1. Calculate the molar weight of an unknown gas if 975 mL of it at 19°C and 0.962 atm weighs 7.29 g.

2. A tea kettle contains a mass of 850. g of water. If all the water is boiled away, what volume would the 850. g of steam occupy at 100.°C and 1.05 atm?

3. Calculate the density of ethane gas (C_2H_6) at 10.°C and 2.00 atm.

4. Why would a gas at low temperature and high pressure deviate from ideal behavior?

Experiment 12   The Determination of the Molar Weight of Butane

5. If the Goodyear blimp holds 5.91 x 10^6 L of helium at 29°C and 775 mmHg, what is the mass of the helium that it contains?

6. A balloon filled with hydrogen is released and floats upward. At some point, the balloon explodes. Explain why these events occurred.

7. If a 17.8 g sample of nitrogen gas at 26°C occupies a volume of 2.88 L, what must be the pressure of the gas?

8. The reaction used to fill an automobile air bag with nitrogen gas is:

$$2\,NaN_3 \rightarrow 2\,Na + 3\,N_2$$

If in order to fill a typical air bag, 150. L of nitrogen at STP are required, how many grams of sodium azide (NaN$_3$) must be used to produce this gas? **Hint: 1 mole of any gas at STP occupies 22.4 L.**

9. A sample of oxygen gas weighs 52.3 g. If it is confined in a container having a volume of 18.6 L and exerts a pressure of 2.15 atm, what must be its temperature in °C?

10. A gas weighs 0.500 g and occupies a volume of 1.00 L at 27 °C and 0.220 atm. Calculate the molar weight of the gas. The molecular formula of this compound is:

    a. CH_2
    b. C_2H_4
    c. C_3H_6
    d. C_4H_8
    e. C_5H_{10}

Experiment 12 The Determination of the Molar Weight of Butane

Name:_____   Section:_____

EXPERIMENT **13**

Chemical Equilibrium and Le Chatelier's Principle

I. OBJECTIVES:

A. To understand dynamic equilibrium and Le Chatelier's Principle.

B. To observe the change of an equilibrium when the concentration of a reactant or product is altered.

C. To be able to predict the affect of concentration changes on chemical equilibria.

II. DISCUSSION:

If water is placed into an Erlenmeyer flask, and the flask is stoppered, a series of molecular changes occur. Initially, the water begins to evaporate. This rate of evaporation depends on the following factors:

1. Temperature
2. The hydrogen bonding attractions of the water molecules for each other.
3. The pressure of the air above the water.
4. The surface area of the water.

Since these factors do not change significantly during the evaporation (as long as the flask is stoppered), the rate of evaporation is <u>constant</u>. This rate is represented in **Figure 1a** as an arrow pointing upward. A small arrow is shown pointing downward, indicating that initially, since very few water molecules have entered the gaseous state, only a relatively small number of them are reentering the liquid state by random motion. After sufficient time has passed, enough water has evaporated so that the number of water molecules in the gaseous state is large enough so that the number of them reentering the liquid state equals the number entering the gaseous state, i.e.,

Number of water molecules $_{\text{(liquid to gas)}}$ = Number of water molecules $_{\text{(gas to liquid)}}$

When this situation occurs, the water has reached a state of **dynamic equilibrium** between its liquid and gaseous states. This is shown in **Figure 1b**, where the arrows representing the movement of water molecules are of equal length.

Experiment 13 Chemical Equilibrium and Le Chatelier's Principle

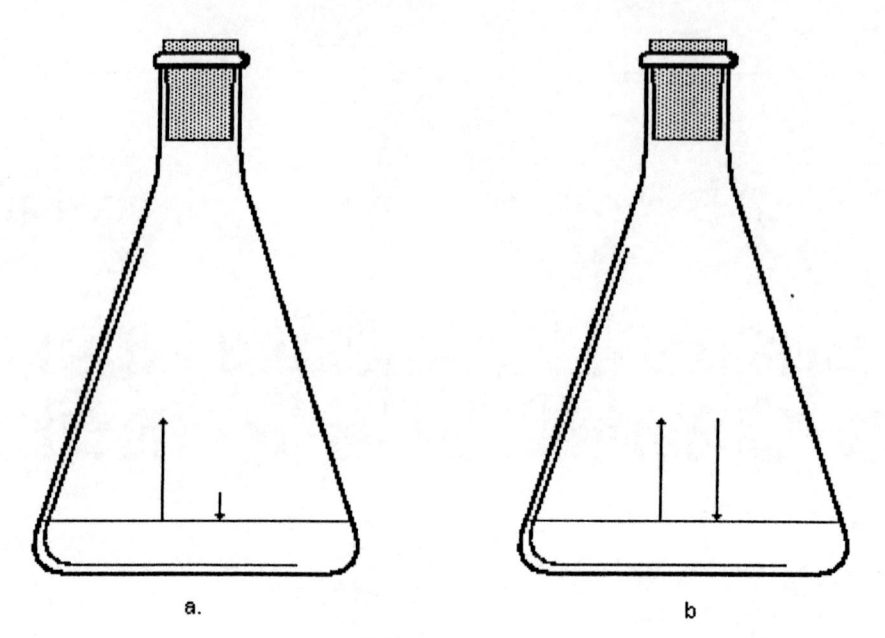

Figure 1

Now, imagine that the water in the flask is warmed slightly. This will increase the kinetic energy of the water molecules in the liquid state, allowing more of them to enter the gaseous state. For a moment, the equilibrium will be disrupted, as shown in **Figure 2a**. Eventually, enough water molecules will have entered the gaseous state, so that by random collisions, an equal number of molecules will return to the liquid state. Once again, a situation of dynamic equilibrium will have been reached, as is shown in **Figure 2b**.

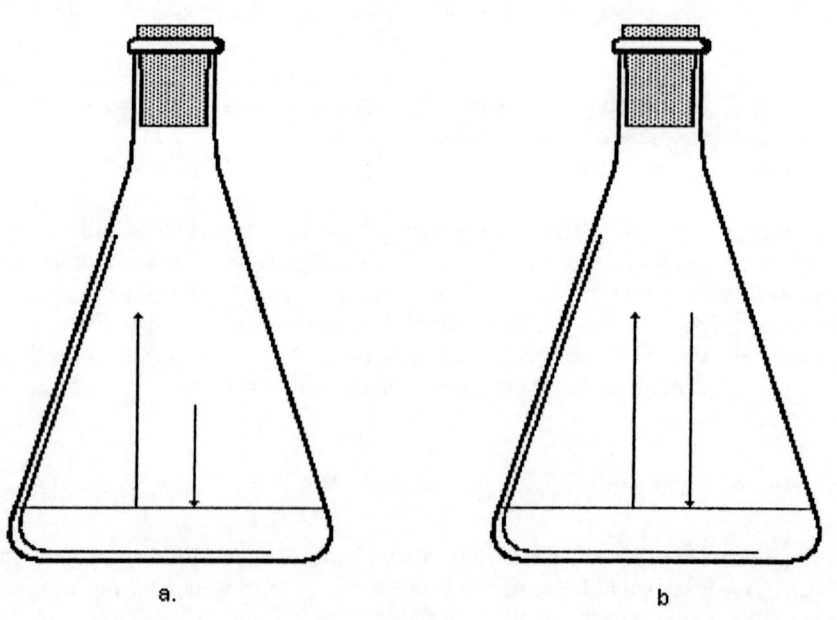

Figure 2

The initial system that was at equilibrium at the lower temperature, responds to the addition of heat by establishing a new equilibrium. This can be described by an equilibrium equation, i.e.,

$$\text{Water}_{(liquid)} + \text{heat} \rightleftarrows \text{Water}_{(gas)}$$

Experiment 13 Chemical Equilibrium and Le Chatelier's Principle

When heat is added, the equilibrium shifts to the right to produce a new equilibrium at higher temperature. This new equilibrium state consists of a larger amount of water vapor, and a smaller amount of liquid water.

Chemical reactions can respond to changes in temperatures or to changes in the concentrations of reactants and products. The effect of these changes was first described by Henri Le Chatelier in 1888. Now known as **Le Chatelier's Principle**, it states that if a variable in a system at equilibrium is changed, the other variables change in the direction that reduces the effect of the change. Thus, the equilibrium reaction:

$$A + B \rightleftharpoons C + D$$

can be shifted rightward by adding more A or B. Likewise, by adding C or D, the equilibrium can be shifted to the left. Additionally, if the concentration of a reactant or product can be reduced in an equilibrium, the reaction will shift toward that side of the reaction.

The equilibrium can be studied quite easily if there is an acid or base on one side of the equation, or if only one side contains an insoluble substance, or if the species on the two sides have different colors.

In the first part of this experiment, you will investigate the equilibrium involved in a saturated salt water solution. This equilibrium can be described as:

$$NaCl(s) \rightleftharpoons Na^+(aq) + Cl^-(aq)$$

By adding concentrated hydrochloride acid, which contains chloride ion, you will disrupt this equilibrium. Since you will have increased the concentration of a product, according to Le Chatelier's Principle, the equilibrium should shift leftward.

A solution of iron(III) ions and thiocyanate ions exist as an equilibrium with thiocyanatoiron(III) ions:

$$Fe^{3+}(aq) + SCN^-(aq) \rightleftharpoons Fe(SCN)^{2+}(aq)$$
$$\text{yellow} \qquad\qquad\qquad\qquad \text{deep red}$$

To this solution, you will add more Fe^{+3} ions and observe any change in the color of the reaction mixture. In a separate test, you will add SCN^{-1} ions, and observe any change in color. Finally, by adding OH^{-1} ions, you will precipitate Fe^{+3} ions as the insoluble $Fe(OH)_3$. Again, you will look for a change in color.

When acetic acid is dissolved in water, the following equilibrium is established:

$$HC_2H_3O_2(aq) + H_2O(l) \rightleftharpoons H_3O^+(aq) + C_2H_3O_2^-(aq)$$

None of the species have any color. However, since the hydronium ion is present, an indicator, methyl orange, can be used to determine its concentration. At high concentrations of H_3O^+, methyl orange has a red color. As the concentration of H_3O^+ decreases, the methyl orange changes to an orange and then to a yellow. You will test Le Chatelier's Principle with this equilibrium by adding sodium acetate, i.e., by increasing the concentration of acetate ions. In a separate test, you will add sodium chloride. A third test will be performed by adding sodium hydroxide solution to the acetic acid solution. Hydroxide ions have a great affinity for hydronium ions, and unite with them to reduce their concentration on the right side of the equilibrium.

Experiment 13 Chemical Equilibrium and Le Chatelier's Principle

In an acidic solution, chromate ion (CrO_4^{-2}) is in equilibrium with dichromate ion ($Cr_2O_7^{-2}$), according to the following equation:

$$2\,CrO_4^{2-}(aq) + 2\,H_3O^+(aq) \rightleftharpoons Cr_2O_7^{2-}(aq) + 3\,H_2O(l)$$

yellow ⟶ orange

You will obtain a solution of potassium chromate and observe its color. Nitric acid will then be added and its effect on the equilibrium will be observed in terms of a color change. The acidity will be reduced by adding sodium hydroxide, and once again a shift in the equilibrium noted via a color change.

Bismuth chloride reacts with water to form an equilibrium with bismuth oxychloride and hydrochloric acid.

$$BiCl_3(aq) + H_2O(l) \rightleftharpoons BiOCl_3(s) + 2\,HCl(aq)$$

You will dissolve bismuth chloride in water and observe the establishment of the equilibrium. The equilibrium will then be shifted to the left by the addition of hydrochloric acid. Then, by adding water, you will be able to shift the equilibrium rightward.

When cobalt(II) chloride is dissolved in water, the cobalt ions exist as complexes with water, i.e., $Co(H_2O)_6^{+2}$. These ions are in equilibrium with a tetrachloro complex, i.e.,

$$Co(H_2O)_6^{2+}(aq) + 4\,Cl^-(aq) \rightleftharpoons CoCl_4^{2-}(aq) + 6\,H_2O(aq)$$

pink ⟶ blue

By varying the concentration of chloride ions and of water molecules, you will be able to shift the equilibrium, and observe the corresponding changes in color.

III. MATERIALS:

A set of 4" and 6" test tubes, test tube rack, pipets, 100 mL graduated cylinder, 250 mL beaker, saturated sodium chloride solution, concentrated hydrochloric acid, 0.1 M iron(III) chloride, 0.1 M potassium thiocyanate, 6 M NaOH, 50% NaOH solution, 0.1 M acetic acid, methyl orange, sodium acetate, sodium chloride, 0.1 M potassium chromate, 6 M nitric acid, bismuth chloride, 1 M cobalt(II) chloride.

IV. PROCEDURE:

As you finish each procedure, dispose of the contents of the test tubes as directed by your instructor, wash the glassware with soap and water, and return the equipment to the designated area.

A. The Saturated Sodium Chloride Solution Equilibrium:

1. To a 4" test tube, add about 5 mL of saturated sodium chloride (**NaCl**) solution. Record its appearance in **Table 1**.

2. To this solution, add several drops of concentrated hydrochloric acid (**HCl**). Record your observations in the Table.

Experiment 13 Chemical Equilibrium and Le Chatelier's Principle

Table 1

	Observations
Saturated NaCl solution	
Saturated NaCl solution plus HCl	

B. The Iron(III) Thiocyanate Ion Equilibrium:

1. To 100 mL of water in a 250 mL beaker, add 2 mL of 0.1 M iron(III) chloride **(FeCl$_3$)** solution and 2 mL of 0.1 M potassium thiocyanate **(KSCN)** solution. Stir this **stock solution** until it is homogeneous. What is its appearance? Record this in **Table 2**.

2. To a 4" test tube **(Tube 1)**, add about 5 mL of the stock solution. To this solution, add about 20 drops of 0.1 M iron(III) chloride **(FeCl$_3$)** solution. Record your observations in the Table.

3. To a 4" test tube **(Tube 2)**, add about 5 mL of the stock solution. To this solution, add about 20 drops of 0.1 M potassium thiocyanate **(KSCN)** solution. Again, record your observations in the Table.

4. To a third 4" test tube **(Tube 3)**, add 5 mL of the stock solution and then 5 drops of 6 M sodium hydroxide **(NaOH)** solution. Record what you see in the Table.

5. To **Tube 2**, add 2 drops of a 50% sodium hydroxide **(NaOH)** solution. Record what you see happening. Then add 5 drops of 12 M hydrochloric acid **(HCl)** and record your observation.

Table 2

	Observations
The Fe(SCN)$^{2+}$ solution	
Tube 1 The Fe(SCN)$^{2+}$ solution + FeCl$_3$ solution	
Tube 2 The Fe(SCN)$^{2+}$ solution + KSCN solution	
Tube 3 The Fe(SCN)$^{2+}$ solution + NaOH solution	
Tube 2 The mixture after the addition of 2 drops of the 50 % NaOH solution	
Tube 2 The mixture after the addition of 5 drops of 12 M HCl solution	

Experiment 13 Chemical Equilibrium and Le Chatelier's Principle

C. The Acetic Acid Equilibrium:

1. To each of three 4" test tubes, add about 3 mL of a 0.1 M acetic acid ($HC_2H_3O_2$) solution. To each of the tubes, add a few drops of methyl orange solution, and agitate the tubes until the solutions are homogeneous. In **Table 3**, record your observations.

2. To the first test tube **(Tube 1)**, add a few crystals of sodium acetate ($NaC_2H_3O_2$). Agitate the tube in order to dissolve the solid. Record your observations in the Table.

3. To the second test tube **(Tube 2)**, add a few crystals of sodium chloride **(NaCl)**. Agitate the tube until the solid dissolves. Record your observations in the Table.

4. To the third test tube **(Tube 3)**, add a few drops of 6 M sodium hydroxide **(NaOH)** solution. Agitate the tube and record your observations.

Table 3

	Observations
Acetic acid solution with methyl orange	
Tube 1 Acetic acid solution + $NaC_2H_3O_2$	
Tube 2 Acetic acid solution + NaCl	
Tube 3 Acetic acid solution + NaOH	

D. The Chromate-Dichromate Equilibrium:

1. To a 4" test tube, add about 5 mL of a 0.1 M potassium chromate (K_2CrO_4) solution. Observe and record its color in **Table 4**.

2. To this solution, add dropwise a 6 M nitric acid (HNO_3) solution, until a distinct change is noted. Record your observations in the Table.

3. Now, to the test tube, add dropwise a 6 M sodium hydroxide **(NaOH)** solution, until once again a distinct change has been observed. Record this in the Table.

Experiment 13 Chemical Equilibrium and Le Chatelier's Principle

Table 4

	Observations
The K_2CrO_4 solution	
The K_2CrO_4 solution + HNO_3	
The K_2CrO_4 solution + NaOH	

E. The Bismuth Chloride-Water Equilibrium:

1. To a 6" test tube, add 2 mL of distilled water. Add a small crystal of bismuth chloride (**$BiCl_3$**) to the water and agitate the tube. Record your observations in **Table 5**.

2. To this mixture, add dropwise 12 M hydrochloric acid (**HCl**) while agitating the test tube, until you effect a distinct change. Record your observation of this change in the Table.

3. Now begin adding water (in dropperful portions), with agitation, to the test tube, until once again you see a distinct change. Record this in the Table.

Table 5

	Observations
Bismuth chloride and water	
Mixture after the addition of HCl	
Mixture after the addition of H_2O	

Experiment 13 Chemical Equilibrium and Le Chatelier's Principle

F. The $Co(H_2O)_6^{2+} - CoCl_4^{2-}$ Equilibrium:

1. To a 4" test tube, add about 5 drops of a 1 M cobalt(II) chloride **(CoCl$_2$)** solution. Note its color and record this observation in **Table 6**.

2. Add 12 M hydrochloric acid **(HCl)** dropwise with agitation until you notice a significant change. Record your observations in the Table.

3. Now add water dropwise, with agitation, and record the change you observe occurring in the test tube.

Table 6

	Observations
The original CoCl$_2$ solution	
The CoCl$_2$ solution + HCl	
The CoCl$_2$ solution diluted with H$_2$O	

4. When you have finished your test, dispose of the contents of the test tubes as directed by your instructor, wash the glassware with soap and water, and return the equipment to the designated area.

Name:_____   Section:_____

Experiment 13 Problems

1. What is meant by a dynamic equilibrium?

2. In your own words, explain Le Chatelier's Principle.

3. Consider the following equilibrium reaction:

$$N_2(g) + 3H_2(g) \rightleftarrows 2NH_3(g) + \text{heat}$$

How would the equilibrium shift if:

   a. more nitrogen gas was added?

   b. some of the ammonia gas was removed?

   c. the temperature of the reaction was decreased?

4. What would occur if a few drops of saturated Na_2SO_4 solution were added to a saturated NaCl solution?

5. Explain the results of adding sodium hydroxide solution to the iron(III) thiocyanate solution in Procedure B4.

6. In Procedure D3, explain the reason for the change you observed.

7. A saturated silver bromide solution exhibits the following equilibrium:

$$AgBr(s) \rightleftarrows Ag^+(aq) + Br^-(aq)$$

If a concentrated hydrobromic acid solution, HBr $_{(aq)}$, was added to this mixture, what would you expect to occur?

8. Nitrogen dioxide reacts with itself to produce dinitrogen tetroxide, according to the following equilibrium:

$$2\,NO_2(g) \rightleftarrows N_2O_4(g) + heat$$

If the temperature of the reaction mixture is increased, what will happen to the concentrations of reactant and product?

Name:_____   Section:_____

EXPERIMENT 14

The pH of Common Substances

I. OBJECTIVES:

A. To understand the nature of an indicator.

B. To extract an indicator from red cabbage.

C. To prepare a series of solutions of known pH by serial dilutions.

D. To determine the color of the cabbage indicator at different pH values.

E. To determine the pH of household substances using the cabbage indicator.

II. DISCUSSION:

Acids and bases are found in almost every aspect of our lives. Some common examples are vitamin C (ascorbic acid), household lye (sodium hydroxide), aspirin (acetyl salicylic acid), DI-GEL (magnesium hydroxide), boric acid, codeine, vinegar (acetic acid), pool acid and stomach acid (hydrochloric acid), battery acid (sulfuric acid), and Alka-Seltzer (a sodium bicarbonate / citric acid mixture).

The strength of an acid solution is usually expressed in terms of the molar concentration of the hydrogen ion, or more exactly, of the hydronium ion, H_3O^+. A neutral solution will have a hydronium ion concentration of exactly 10^{-7} M. If the hydronium ion concentration is greater than this value, the solution is acidic, while if it is less than this value, it is basic. A more convenient way of expressing acidity is by the use of the pH scale. The pH is defined as the $-\log[H^+]$. Thus, a neutral solution has a pH of exactly 7, and acidic solution has a pH of less than 7, and a basic solution has a pH of greater than 7.

A convenient way to determine whether a solution is acidic or basic is by the use of an indicator. Indicators are weak acids or bases, which in solution are in equilibrium with their conjugate base or acid. For example, if **HIn** represents a weak acid, its conjugate base is **In⁻**. The equilibrium between the two is:

$$HIn(aq) \rightleftarrows H^+(aq) + In^-(aq)$$

The acid and its conjugate base have different colors. For example, a water solution of phenolphthalein undergoes the following equilibrium:

Experiment 14 The pH of Common Substances

$$\text{HIn (colorless)} \rightleftharpoons H^+ + \text{In}^- \text{ (red)}$$

If phenolphthalein is added to an acidic solution, its conjugate base is protonated, and the colorless form predominates. If it is added to a basic solution, the hydrogen ions react with hydroxide ions, and the equilibrium shifts rightward to produce the red phenolphthalein anion. Some other common indicators used in the lab are:

Indicator	Acid Form	Base Form
Litmus	red	blue
Bromothymol blue	yellow	blue
Phenol red	yellow	red
Methyl orange	red	yellow

In this experiment, you will extract a group of compounds called anthocyanins from red cabbage. This extract will then be mixed with solutions of known pH to determine the color of the extract at different acid concentrations. The extract will then be used to test a variety of household substances. Based on the color change of the cabbage extract, the pH of these solutions will be determined.

III. MATERIALS:

Preboiled distilled water, shredded red cabbage, blender, strainer, 14 250 mL beakers, 10 mL and 100 mL graduated cylinders, labels, 13 4" test tubes, test tube rack, 0.1 M hydrochloric acid, 0.1 M sodium hydroxide, lemon juice, vinegar, baking soda, lye soap, liquid dishwashing soap, aspirin, ammonia solution, battery acid, carbonated soda, salt, sugar.

IV. PROCEDURE:

A. The Extraction of Red Cabbage:

1. Place enough shredded red cabbage in a 250 mL beaker until it is about one-fourth full. Add 100 mL of distilled water.

2. Pour the contents into a blender and blend on high speed for 30 seconds. Strain the blend and collect the purple extract in your beaker.
(Alternately, place the beaker on a piece of wire gauze on an iron ring and heat the mixture to boiling with a Bunsen burner. Continue boiling the water until it is a deep purple color. Turn off the burner and continue with the next part of the experiment while the extract cools.)

Experiment 14 The pH of Common Substances

B. The Preparation of Different pH Solutions by Serial Dilutions:

1. Label thirteen clean, dry 250 mL beakers from 1 to 13.

2. To beaker 1 add 100 mL of a 0.1 M hydrochloric acid (**HCl**) solution. Since this solution will have a hydrogen ion concentration of 10^{-1} M, it has a pH of 1.

3. By using a 10 mL graduated cylinder, add 10 mL of this solution to a clean 100 mL graduated cylinder. Dilute this solution with 90 mL of preboiled distilled water. This will give you a solution that has a pH of 2. Place this solution in beaker 2. From now on, use only preboiled water in this experiment.

4. Add 10 mL of this solution to a clean 100 mL graduated cylinder (make sure it has been rinsed with distilled water) and then dilute it with 90 mL of distilled water. This will produce a solution of pH 3. Place this in beaker 3.

5. Continue this process of serial dilutions until you have made solutions with pH's of 1-6.

6. In beaker 7 put 100 mL of preboiled distilled water. This will have a pH of 7.

7. To beaker 13 add 100 mL of a 0.1 M sodium hydroxide (**NaOH**) solution. This will give you a solution of pH 13.

8. Add 10 mL of this solution to a clean 100 mL graduated cylinder, and then dilute it with 90 mL of distilled water. This will produce a solution of pH 12. Place this solution in beaker 12.

9. Repeat this process of serial dilutions until you have made solutions with pH values of 8-13.

C. The Determination of the Color of Red Cabbage Extract at Different pH Values:

1. Label a series of 13 4" test tubes from 1 to 13 and place them in a test tube rack.

2. To each test tube, add about 10 mL of the solution of known pH that you prepared in Part B.

3. To each solution, add about 1 mL of the cabbage extract that you prepared in Part A, and record your observations below.

Table 1

pH	Color	pH	Color	pH	Color
1		6		11	
2		7		12	
3		8		13	
4		9			
5		10			

4. Rinse out all of your test tubes thoroughly with distilled water for the next part of the experiment.

Experiment 14 The pH of Common Substances

D. The Determination of the pH of Household Substances:

1. To a series of labeled test tubes, add about 10 mL of the following solutions: lemon juice, vinegar, baking soda, lye, soap, aspirin, ammonia solution, battery acid, a carbonated beverage, salt, sugar.

2. Add several milliliters of the cabbage extract to each of these test tubes and note the color of the resulting solutions. Record your observations in **Table 2**. Based on your results in Part C, estimate the pH of each of these solutions.

Table 2

Solution	Color	pH
lemon juice		
vinegar		
baking soda		
lye		
soap		
aspirin		
ammonia solution		
battery acid		
a carbonated beverage		
salt		
sugar		

Name:_____   Section:_____

Experiment 14 Problems

1. It would be difficult to determine the pH of blood by using cabbage extract. Why?

2. Why do you think that the distilled water you used in this experiment had to be boiled before use? Hint: A gas could be dissolved in water that might affect its pH.

3. If the distilled water was not boiled, what color would this water have produced when treated with the cabbage extract?

4. What are the pH's of solutions with the following hydrogen ion concentrations?

   a. $[H^+] = 10^{-3}$ M

   b. $[H^+] = 10^{-6}$ M

   c. $[H^+] = 10^{-11}$ M

4. What are the hydrogen ion concentrations of solutions whose pH's are:

   a. 12

   b. 7

   c. 1

Experiment 14 The pH of Common Substances

5. Arrange the acidic substances you tested in **Part D** from the strongest acid to the weakest acid.

_____ (strongest acid)

_____

_____

_____

_____ (weakest acid)

6. Arrange the basic substances you tested in **Part D** from the strongest base to the weakest base.

_____ (strongest base)

_____

_____

_____

_____ (weakest base)

Name:_____   Section:_____

EXPERIMENT 15

Buffers and pH Changes

I. OBJECTIVES:

A. To understand the nature of a buffer.

B. To prepare a buffer from acetic acid and sodium acetate.

C. To test the ability of buffered and unbuffered solutions to resist pH changes when strong acids and bases are added.

II. DISCUSSION:

In a healthy individual, the pH of the blood falls in the range of 7.35-7.45. Even when we eat very acidic substances like oranges, tomatoes, and vinegar, the blood's pH does not decrease greatly. Likewise, vomiting results in the loss of stomach acid (hydrochloric acid) and yet the pH of the blood is maintained at a relatively steady level. How is the body able to maintain the pH of the blood so effectively?

The answer is buffers. A buffer solution is a solution that is able to resist pH changes when small amounts of strong acids or bases are added. A buffer can be prepared in two ways.

    1. By combining a weak acid and its salt.
        Ex. $HC_2H_3O_2$ and $NaC_2H_3O_2$
                 H_2CO_3 and $NaHCO_3$

    2. By combining a weak base and its salt.
        Ex. NH_3 and NH_4Cl

In this experiment, you will prepare buffer solutions of acetic acid and sodium acetate. To this, you will add hydrochloric acid and sodium hydroxide solutions and observe the buffering action, i.e., the resistance to pH changes.

What is the chemistry that lies behind this buffering action? A key ingredient is that the solution contains a relatively high concentration of acetic acid molecules <u>and</u> acetate ions. When a strong acid is added to this buffer, it reacts with the acetate ions to produce acetic acid and water.

$$H_3O^+ + C_2H_3O_2^- \rightleftarrows HC_2H_3O_2 + H_2O$$

Experiment 15 Buffers and pH Changes

Note that this reaction lies far to the right and is simply the reverse of the reaction that occurs when acetic acid is added to water. Thus, when strong acid is added and most of the hydronium ions are consumed, the pH will change very little.

If base is added to the buffer, it reacts with the acetic acid molecules to give acetate ions and water.

$$OH^- + HC_2H_3O_2 \rightleftharpoons H_2O + C_2H_3O_2^-$$

Again, since the reaction lies far to the right, most of the hydroxide ions are consumed, and the pH remains relatively constant.

If excessive amounts of either acid or base are added, they can overwhelm the buffering capacity of the solution, and the pH will change considerably. Even in our blood, the buffering system can be overcome occasionally. This buffer is principally due to H_2CO_3/HCO_3^-, with the addition of the participation of carbon dioxide. These substances are linked through two equilibria:

$$H_2O + CO_2 \rightleftharpoons H_2CO_3 \rightleftharpoons H^+ + HCO_3^-$$

With a mild increase in acid in our blood, the buffer system responds by consuming the excess acid with bicarbonate ion. The equilibria shift leftward, with the result that the acid has become incorporated in the water molecules. If on the other hand, there is a mild increase in base, it reacts with the hydrogen ions and the equilibria shift rightward. The result is that the depleted acid is replaced by the reaction of water with carbon dioxide.

Very rapid or heavy breathing (hyperventilation) can cause severe loss of carbon dioxide. This causes a shift in the equilibria leftward, with a resulting increase in the blood's pH. This is termed <u>respiratory alkalosis</u>. Severely diminished respiration (hypoventilation) can lead to a buildup of carbon dioxide in the blood. This shifts the equilibria rightward and the pH of the blood can decrease considerably. This is termed <u>respiratory acidosis</u>.

III. MATERIALS:

10 100 mL beakers, labels, 100 mL graduated cylinder, balance, stirring rod, spatula, pH meter, 1 mL pipet, magnetic stirrer, spin bar, pH paper, 0.10 M sodium chloride, 0.10 M acetic acid, 6 M hydrochloric acid, 6 M sodium hydroxide, sodium acetate.

IV. PROCEDURE:

A. The Preparation of the Buffer Solution:

1. Obtain ten 100 mL beakers and label them 1-10.

2. Add 50 mL of distilled water to beakers 1 and 6.

3. Add 50 mL of 0.10 M sodium chloride (**NaCl**) solution to beakers 2 and 7.

4. To beakers 3 and 8 add 1.00 g of solid sodium acetate (**NaC_2H_3O_2**).

5. To beakers 4 and 9 add 5.00 g of solid sodium acetate.

6. To beakers 5 and 10 add 10.00 g of solid sodium acetate.

Experiment 15 Buffers and pH Changes

7. Add 50 mL of 0.10 M acetic acid (**HC$_2$H$_3$O$_2$**) to each of the beakers 3,4, and 5. Stir each mixture with a clean stirring rod until the sodium acetate has completely dissolved.

B. The Determination of Buffering Action Toward Acid:

1. Obtain about 50 mL of 6 M hydrochloric acid (**HCl**) and a clean 1 mL pipet.

2. Obtain a pH meter, magnetic stirrer, and spin bar. If these are not available, obtain a roll of pH paper.

3. After adjusting your pH meter as directed by your instructor, rinse the probe with distilled water and insert it into beaker 1 containing the distilled water.

4. Set the instrument to read the pH and record your observed value in **Table 1** below.

5. By use of the pipet, add 1 mL of 6 M hydrochloric acid, and record the new pH.

6. Continue adding 1 mL portions of 6 M hydrochloric acid and recording the pH, until the value seems to change only slightly.

7. Turn the pH meter to standby, remove the probe, wash it with distilled water, and replace beaker 1 with beaker 2. Repeat the process described in Steps 3-6.

8. Repeat the pH determinations with the solutions in beakers 3-5.

Table 1

Beaker	1	2	3	4	5
mL of HCl	pH	pH	pH	pH	pH
0					
1					
2					
3					
4					
5					
6					
7					
8					
9					
10					

Experiment 15 Buffers and pH Changes

C. The Determination of Buffering Action Toward Base:

1. Obtain about 50 mL of 6 M sodium hydroxide **(NaOH)**.
2. Rinse out your pipet with distilled water, and then with the sodium hydroxide solution.
3. Repeat the procedures you carried out in Section B, but this time add the sodium hydroxide solution. Record your results in **Table 2**.

Table 2

Beaker	6	7	8	9	10
mL of NaOH	pH	pH	pH	pH	pH
0					
1					
2					
3					
4					
5					
6					
7					
8					
9					
10					

4. When you are finished with your measurements, clean your glassware with tap water and then rinse with distilled water. Return the equipment to the designated area.

Experiment 15 Buffers and pH Changes

Name:_____   Section:_____

Experiment 15 Problems

1. Which is more acidic: a solution with pH 2 or a solution with pH 12?

2. What chemical action does a buffer perform?

3. What are the major characteristics of a buffer solution?
    a.

    b.

    c.

    d.

4. What chemical species would be present in a buffer that contained citric acid ($H_3C_6H_5O_7$) ?

5. A treatment for respiratory alkalosis is to breathe into a paper bag. Explain what happens chemically during this treatment.

6. From your results in this experiment, which solution of those you tested had the greatest buffer capacity:

    a. toward strong acid?

    b. toward strong base?

Experiment 15 Buffers and pH Changes

7. If sodium dihydrogen phosphate is added to sodium bicarbonate, the following reaction occurs:

$$NaH_2PO_4 + NaHCO_3 \rightleftarrows Na_2HPO_4 + H_2CO_3$$

a. Which reactant is behaving as the acid? _____

b. Which product is behaving as the base? _____

8. Why was distilled water used to rinse off the pH probe?

9. A buffer solution is needed in the lab. Pick out pairs of chemicals from the list below, which when mixed together in water, would produce a buffer solution.

HCl	NH_3	Na_3PO_4	$HC_2H_3O_2$	KNO_3
$MgCl_2$	NaCl	$KC_2H_3O_2$	NH_4NO_3	H_2CO_3
HNO_3	$NaHCO_3$	NaOH	HBr	NH_4Cl

a. _____ and _____

b. _____ and _____

c. _____ and _____

10. What is the $[H^+]$ of a solution with a pH equal to:

a. 2

b. 4

d. 11

11. What is the pH of a hydrochloric acid solution that has a $[H^+]$ of 1×10^{-5} ?

12. What is the pH of an acetic acid solution that has a $[H^+]$ of 1×10^{-6} ?

Name:_____   Section:_____

EXPERIMENT 16

Electrolysis and Electroplating

I. OBJECTIVES:

A. To understand the chemical changes occurring during the electrolysis of water solutions of ionic compounds.

B. To electrolyze water and some aqueous solutions and observe the resulting chemical changes.

C. To understand the process of electroplating.

D. To electroplate a piece of copper with zinc.

II. DISCUSSION:

Electrochemical reactions involve the transfer of electrons to bring about a chemical change. When this process is nonspontaneous, i.e., when electricity has to be supplied to cause the reaction to occur, it is termed **electrolysis**. An example of a nonspontaneous reaction is the decomposition of water to hydrogen and oxygen.

When electricity is passed through water, two chemical changes, called half-reactions, occur. At the negative electrode (**cathode**) electrons are released to the water molecules, resulting in its conversion to hydrogen molecules and hydroxide ions.

$$2H_2O + 4e^- \rightarrow 2H_2 + 4OH^-$$

The gaining of electrons by water (or other substances) is termed **reduction**.

At the positive electrode (**anode**), electrons are released from the water to the electrode. At this electrode, water molecules are converted to oxygen molecules, hydrogen ions, and electrons (which go into the electrode to complete the electrical circuit).

$$2H_2O \rightarrow O_2 + 4H^+ + 4e^-$$

The loss of electrons by water (or other substances) is termed **oxidation**.

These two processes constitute the electrolytic cell shown in **Figure 1**.

Experiment 16 Electrolysis and Electroplating

$$4H_2O + 4e^- \longrightarrow 2H_2 + 4OH^- \qquad 2H_2O \longrightarrow O_2 + 4H^+ + 4e^-$$

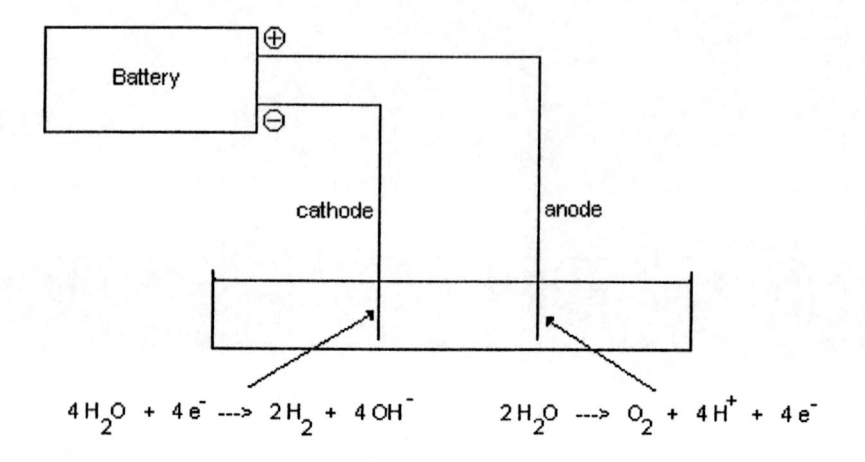

Figure 1

Since hydrogen ions and hydroxide ions have an extremely high affinity for each other, they combine to give water molecules. If these two reactions are added together, and the excess water molecules subtracted from each side of the equation, the result is:

$$2H_2O \rightarrow 2H_2 + O_2$$

In this experiment, you will pass electricity through water and observe the formation of gases at each electrode. By adding an indicator to the solution, you will be able to determine which half-reaction is occurring at each electrode.

If other substances are present that can undergo electrolysis more easily than water, they will preferentially be oxidized or reduced. You will test this by electrolyzing a variety of aqueous solutions.

A variation of electrolysis is **electroplating**. In this process, an electrolytic cell consists of two metal electrodes. One of the metals can be deposited on the surface of the other metal when electricity is passed through a solution containing ions of the first metal. Electroplating is used to chrome plate automobile parts, silver plate eating utensils, and zinc plate (galvanize) metal garbage cans. In this experiment, you will electroplate a strip of copper metal with zinc.

III. MATERIALS:

9 V battery, Petri dish, two sets of insulated copper wires with alligator clips at each end, two graphite electrodes, sodium sulfate, stirring rod, bromthymol blue indicator solution, universal indicator solution, sodium iodide solution, sodium chloride solution, starch solution, copper(II) sulfate solution, copper strip, zinc strip, zinc sulfate solution.

IV. PROCEDURE:

A. The Electrolysis of Water:

1. Connect one terminal of a 9 volt battery to an alligator clip on a piece of copper wire. To the other alligator clip of the wire, attach a small piece of graphite. Repeat this process with the other terminal of the battery to give the electrolysis apparatus shown in **Figure 2**.

Experiment 16 Electrolysis and Electroplating

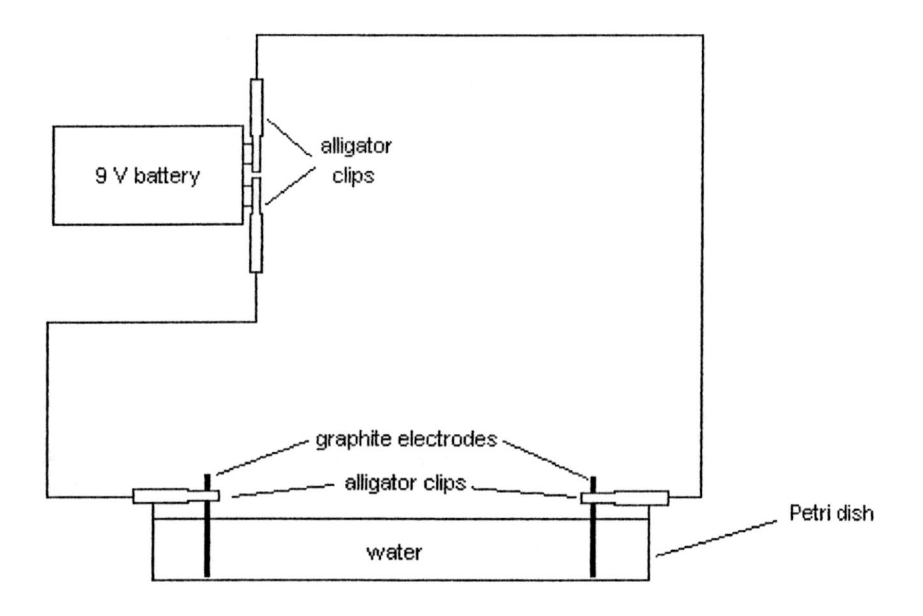

Figure 2

2. To a Petri dish, add distilled water until it is about half full. Then add a few crystals of sodium sulfate (**Na₂SO₄**). The sodium sulfate will allow the water to conduct electricity. Stir the mixture in the dish to help dissolve the sodium sulfate.

3. Immerse the electrodes in the water as shown in Figure 2, and observe what happens at each electrode. Record your observations in **Table 1**.

4. Remove the electrodes from the solution, stir the solution, and add a few drops of universal indicator solution. In basic solutions, universal indicator is purple, in neutral solutions it is green, while in acidic solutions it is red. What is the color of the solution? Record this color in the Table.

5. Again, immerse the electrodes in the solution and observe the changes that occur. Record what you see in the Table.

From your observations, determine which electrode was the cathode and which was the anode. Record your conclusion in **Figure 3**.

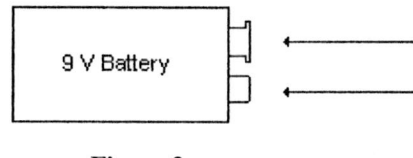

Figure 3

6. Dispose of the solution in the sink. Thoroughly rinse the Petri dish and electrodes with distilled water.

Experiment 16 Electrolysis and Electroplating

Table 1

	Observations
Change occurring at the electrodes during electrolysis	
Color of the solution with universal indicator	
Color of universal indicator at the electrodes during electrolysis	

B. The Electrolysis of a Sodium Iodide Solution:

1. Set up the electrolysis apparatus shown in Figure 2. To the Petri dish, add 0.1 M sodium iodide (**NaI**) solution.

2. Immerse the electrodes in the solution and record any observations you make in **Table 2**.

3. Remove the electrodes. Dispose of the solution, and thoroughly rinse the electrodes and dish. To the dish, add fresh sodium iodide solution, and then a few drops of starch solution. The starch is used to test for the presence of iodine (I_2). Starch and iodine unite to form a complex with a very intense color.

4. Stir the mixture, and then immerse the electrodes. Record your observations in the Table.

5. Dispose of this solution and thoroughly rinse the electrodes and dish with distilled water.

6. Again, add the sodium iodide solution to the dish, followed by a few drops of universal indicator solution. Stir the mixture and immerse the electrodes. Record any changes in the Table.

7. Wash your apparatus as before with distilled water.

Table 2

	Observations
Change occurring at the electrodes during electrolysis	
Change occurring at the electrodes in the solution containing the starch solution	
Change occurring at the electrodes in the solution containing universal indicator	

Experiment 16 Electrolysis and Electroplating

C. The Electrolysis of a Sodium Chloride Solution:

1. Set up the electrolysis apparatus shown in Figure 2. To the Petri dish, add 0.1 M sodium chloride **(NaCl)** solution.

2. Immerse the electrodes in the solution and record your observations in **Table 3**. Note the odor of the gases being evolved.

3. Remove and rinse the electrodes with distilled water. Dispose of the salt solution, and rinse the Petri dish with water.

4. To the dish, add more sodium chloride solution and a few drops of universal indicator solution. Stir the mixture. Immerse the electrodes in the solution and note any changes. Record these observations in the Table.

5. Thoroughly clean your apparatus with distilled water.

Table 3

	Observations
Change occurring at the electrodes during electrolysis	
Changes occurring at the electrodes in the solution containing the universal indicator	

D. The Electrolysis of a Copper(II) Sulfate Solution:

1. Set up the electrolysis apparatus shown in Figure 2. To the Petri dish, add 0.1 M copper(II) sulfate **($CuSO_4$)** solution.

2. Immerse the electrodes in the solution and note any changes that occur at the electrodes. Record what you observe in **Table 4**.

3. Remove the electrodes, wash the apparatus, and pour fresh copper(II) sulfate solution into the dish. Add a few drops of bromthymol blue indicator, and stir the solution. Bromthymol blue indicator is yellow in acidic solution and blue in basic solution.

4. Immerse the electrodes and observe any changes that occur. Record what you see happening in the Table.

5. Clean your equipment thoroughly with distilled water.

Experiment 16 Electrolysis and Electroplating

Table 4

	Observations
Change occurring at the electrodes during electrolysis	
Changes occurring at the electrodes in the solution containing the bromthymol blue	

E. Electroplating Copper with Zinc:

1. Set up the apparatus as shown in Figure 2. However, instead of the graphite electrodes, attach a small piece of copper to the alligator clip that is connected to the cathode. To the alligator clip that is connected to the anode, attach a small piece of zinc.

2. To the Petri dish, add 0.1 M zinc sulfate **($ZnSO_4$)** solution.

3. Immerse the two metal electrodes in the solution and observe any changes that occur. Record what you see in **Table 5**. Dispose of the zinc sulfate solution as directed by the instructor and clean your equipment with distilled water.

4. To the Petri dish, add 0.1 M copper(II) sulfate **($CuSO_4$)**. Attach the alligator clip that is the cathode to the piece of zinc, and the electrode that is the anode to the piece of copper. Immerse the electrodes in the solution and observe any changes that occur. Record these in the Table. Dispose of the copper sulfate as directed by your instructor. Thoroughly clean your equipment with distilled water, and return it to the area designated by the instructor.

Table 5

	Observations
Changes occurring at the electrodes in the zinc sulfate solution	
Changes occurring at the electrodes in the copper(II) sulfate solution	

Name:_____   Section:_____

Experiment 16 Problems

1. a. When you electrolyzed water, what half-reaction occurred at the anode?

   b. What evidence do you have that this reaction took place?

   c. What half-reaction occurred at the cathode?

   d. What is your evidence?

2. Write the half-reaction occurring during the electrolysis of sodium iodide solution at the:

   a. cathode

   b. anode

3. Write the half-reaction occurring during the electrolysis of sodium chloride solution at the:

   a. cathode

   b. anode

4. Write the half-reaction occurring during the electrolysis of copper(II) sulfate solution at the:

   a. cathode

   b. anode

5. Write the half-reaction occurring during the electroplating of copper with zinc at the:

   a. cathode

   b. anode

6. Two possible oxidations could have occurred in the electroplating experiment. They are:

$$Cu(s) \rightarrow Cu^{2+}(aq) + 2e^-$$

$$Zn(s) \rightarrow Zn^{2+}(aq) + 2e^-$$

In aqueous solution, Cu^{+2} ions are blue in color. From your observations, which of these oxidations actually occurred?

7. a. Why was sodium sulfate added to the water before it was electrolyzed?

   b. Why could you not use potassium bromide for this purpose?

Experiment 16 Electrolysis and Electroplating

8. Identify the following half-reactions as oxidation or reduction.

   a. $2H^+ + 2e^- \rightarrow H_2$

   b. $2I^- \rightarrow I_2 + 2e^-$

   c. $Zn^{2+} + 2e^- \rightarrow Zn$

9. a. When water is electrolyzed, is an acid or base produced at the cathode?

   b. How do you know?

10. You tested NaCl and NaI in electrolysis reactions. What would you see at each electrode if you tested NaBr? Hint: Br_2 is reddish-brown in color.

	Cathode	Anode
NaBr solution		

11. a. In electroplating, the goal is to put a layer of a metal on an object. Is the object the cathode or the anode?

   b. Why?

12. Hearing aids and watches are powered by mercury batteries. The overall reaction is:

$$HgO + Zn \rightarrow ZnO + Hg$$

   a. What species is oxidized?

   b. What is the oxidizing agent?

   c. Write the two half-reactions involved in this process.

Experiment 16 Electrolysis and Electroplating

Name:_____   Section:_____

EXPERIMENT 17

The Synthesis and Recrystallization of Aspirin

I. OBJECTIVES:

A. To synthesize aspirin.

B. To collect aspirin by the use of vacuum filtration.

C. To calculate the percentage yield of a reaction.

II. DISCUSSION:

The most common medicinal drug used in the United States is aspirin, acetylsalicylic acid. Every year, approximately nine billion grams of aspirin are consumed. This equates to about 30 billion tablets per year. The popularity of this drug is a combination of its ability to act as a powerful antipyretic (fever reducer), anti-inflammatory agent (swelling reducer), and analgesic (pain reducer), while having relatively few side effects.

Following about 100 years of investigations of pain and fever remedies, it was recognized that the extracts of the meadowsweet plant and the bark of the willow tree had potent medicinal properties. The active ingredient was found to be salicylic acid (from *salix*, the Latin term for the willow tree).

salicylic acid

This compound was found to be quite acidic and caused discomfort to the mouth and upper digestive tract. To alleviate this, salicylic acid could be converted to its salt, sodium salicylate, which is less caustic.

Experiment 17 The Synthesis and Recrystallization of Aspirin

sodium salicylate

This compound possessed an unpalatable sweet taste that made it unsuitable for use by the general public. By the late 1800's it was found that salicylic acid could be converted to acetylsalicylic acid.

acetylsalicylic acid

When it reaches the small intestine, the strongly basic conditions hydrolyze this compound to the salicylate anion, which can then be absorbed into the blood stream. The tradename aspirin, came from *a* for acetyl, and *-spir* from spirea, the Latin term for the meadowsweet plant. While aspirin originally was prepared by the reaction of salicylic acid and acetic acid, this reaction is quiet slow and has been replaced by a more efficient method. When salicylic acid is mixed with an excess of acetic anhydride in the presence of a small amount of sulfuric acid (a catalyst), and the mixture is heated, an equilibrium reaction occurs to produce aspirin.

salicylic acid + acetic anhydride ⇌ acetylsalicylic acid + acetic acid

By using excess acetic anhydride, this reaction is shifted to the right, and increases the yield of aspirin. The product is insoluble in water. Thus, if you add water to your reaction product, the aspirin will precipitate and may be collected by filtration. By using a vacuum to help draw the aqueous solution through the filter paper, you will be able to filter the mixture in a few seconds while it remains cold. This will minimize the loss of product due to dissolution by the water. The vacuum filtration will also remove the bulk of the water. Any residual water will be removed by drying the product in an oven. The product will then be weighed. Due to the equilibrium nature of this reaction, and the possible formation of side-products, the mass of dry product that you obtain will not be the maximum mass that is possible (the theoretical yield). The percentage yield is an indication of the efficiency of the reaction to produce the desired product (along with an indication of the skill of the chemist to minimize mechanical losses). It is the ratio of what you obtained (the actual yield) to what you theoretically could obtain (the theoretical yield), as shown in the following equation.

$$\text{Percentage Yield} = \frac{\text{Actual Yield}}{\text{Theoretical Yield}} \times 100\%$$

After calculating the theoretical yield, you will be able to determine your percentage yield from your actual yield.

III. MATERIALS:

125 mL Erlenmeyer flask with cork, salicylic acid, acetic anhydride, conc. sulfuric acid, balance, 250 mL and 400 mL beakers, wire gauze, iron ring, Bunsen burner, buret clamp, 10 mL and 100 mL graduated cylinders, ice, small Buchner funnel, filter paper, Filtervac, 125 mL filtering flask, vacuum tubing, aspirator, ethanol, stirring rod, watch glass, vial, drying oven.

IV. PROCEDURE:

A. The Synthesis of Aspirin:

1. Tare a clean, dry 125 mL Erlenmeyer flask and record its mass in **Table 1**.

2. To the flask add about 6 g of salicylic acid. Reweigh the flask and record the mass in the Table.

3. Subtract the tare weight from the weight of your flask plus salicylic acid, in order to obtain the weight of the salicylic acid. Record this in the Table.

4. Put about 200 mL of water in a 400 mL beaker. Place the beaker on a piece of wire gauze supported by an iron ring. Heat the water to boiling with a Bunsen burner.

5. While the water is heating, add 10 mL of acetic anhydride to the salicylic acid.

6. To this mixture, add 10 drops of concentrated sulfuric acid (H_2SO_4). Both the acetic anhydride and the sulfuric acid can cause skin burns. If you get any of these chemicals on your hands, immediately wash them thoroughly with water.

7. Loosely cork the Erlenmeyer flask and swirl it to mix the reactants.

8. Clamp the flask in the beaker of boiling water and heat the mixture for about 15 minutes.

9. During this time, all the salicylic acid should dissolve and react.

10. Remove the flask from the beaker and place it under running tap water until it has cooled to near room temperature. Save the beaker of hot water for later use.

11. Add about 20 mL of ice water to the flask. This will decompose the excess acetic anhydride to acetic acid, and also promote the crystallization of the aspirin.

12. Put the flask in a beaker of ice water to maximize the crystallization of the aspirin. Continue cooling the mixture while you set up a vacuum filtration apparatus.

13. Construct a vacuum filtration apparatus as shown in **Figure 1**.

Experiment 17 The Synthesis and Recrystallization of Aspirin

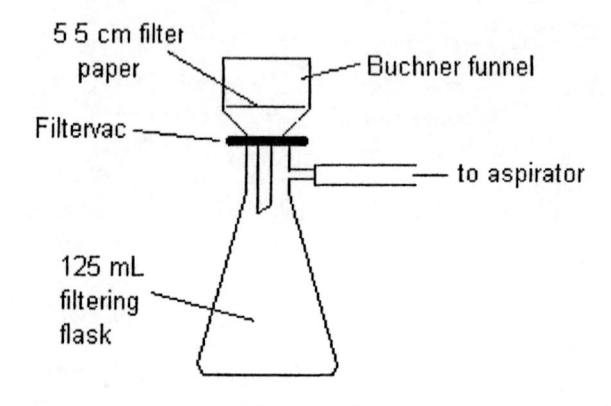

```
        5 5 cm filter
           paper          ┌─────┐
                          │     │───── Buchner funnel
         Filtervac ───────┤     │
                          └──┬──┘
                             │      ───── to aspirator
         125 mL
         filtering
          flask
```

Figure 1

14. Turn on the aspirator all the way. Wet the filter paper with water so that it lays flat on the funnel. Pressing down on the funnel will help to seal the funnel to the Fltervac and the Filtervac to the flask.

15. Pour the ice cold mixture of aspirin into the funnel while continuing to press down on the funnel. Rinse out any solid that remains in the flask by using a small amount of ice cold water.

16. Continue to draw air through the aspirin until no more water appears to be dripping from the funnel.

17. Turn off the aspirator and disassemble the filtration apparatus.

B. The Recrystallization of the Aspirin:

1. Transfer the slightly moist crystals of aspirin to a 150 mL beaker. Add 30 mL of ethanol to this solid.

2. Place the beaker in the hot water bath and swirl until all of the solid dissolves.

3. Obtain about 60 mL of warm water, add this to the aspirin solution, and stir the resulting mixture.

4. Remove the beaker from the hot water bath, cover it with a watch glass, and allow it to cool.

5. While it is cooling, clean the Buchner funnel and filtering flask and once again set up a vacuum filtration apparatus.

6. Continue to cool the mixture in an ice bath until the crystallization appears to be complete.

7. Collect the purified aspirin by vacuum filtration as before. Wash the crystals with a small amount of ice cold water. Remove as much water as possible by drawing air through the crystals.

8. Scrape the crystals onto a watch glass and dry them in an oven that is at 90-100°C.

9. When they appear to be dry, transfer the crystals to a tared vial and reweigh the vial in order to determine the yield of your product. Record all the masses in the Table.

10. Label your vial with your name and turn it in to your instructor.

11. Wash all your equipment with soap and water and return it to the designated area.

Table 1

	Data
Tare weight of the Erlenmeyer flask	g
Mass of flask and salicylic acid	g
Mass of salicylic acid	g
Mass of the empty vial	g
Mass of vial and aspirin	g
Mass of aspirin	g

Experiment 17 The Synthesis and Recrystallization of Aspirin

Experiment 17 The Synthesis and Recrystallization of Aspirin

Name:_____   Section:_____

Experiment 17 Problems

1. From the number of grams of salicylic acid that you started with, calculate the theoretical yield of aspirin that you could obtain. Assume that you used an excess of acetic anhydride.

2. Calculate the percentage yield of aspirin in your experiment.

3. If a reaction yielded 9.69 g of acetylsalicylic acid, what would be the percentage yield of the reaction if the theoretical yield was 11.22g?

4. What is an analgesic?

5. What is an antipyretic?

6. Write the reaction that occurs with aspirin when it enters the small intestine.

Experiment 17 The Synthesis and Recrystallization of Aspirin

7. Why did you add sulfuric acid to your reaction mixture?

8. What was the purpose of recrystallizing your aspirin?

9. a. If your percentage yield was found to be greater than 100 %, what would be the most likely reason? Assume that your calculations are correct, and that you did not make any weighing errors.

   b. What would you do experimentally to correct this?

10. Explain why actual yields are generally lower than theoretical yields.

11. Most aspirin tablets contain 5 grains of aspirin. If 1 grain = 0.0648 g, how many grams of aspirin are there in a tablet?

12. How many aspirin tablets could be made from the aspirin you synthesized in this experiment. Assume that your product was pure aspirin.

Name:_____   Section:_____

EXPERIMENT 18

The Extraction of Caffeine from Coffee

I. OBJECTIVES:

A. To carry out an extraction of an aqueous solution of coffee with an organic solution.

B. To isolate caffeine from coffee.

C. To determine the melting point of caffeine.

II. DISCUSSION:

Every day, throughout the world, people start out their morning by extracting alkaloids from naturally occurring materials, and then drinking this extract. The most common of these alkaloids is 1,3,7-trimethylxanthine, or simply caffeine. Caffeine is found in nature not only in coffee beans and tea leaves, but in cocoa, chocolate, and kola nuts. Additionally, it is used as an additive in a variety of soft drinks such as Coca-Cola®, Pepsi-Cola®, Tab®, and Mountain Dew®. Other common sources of caffeine are stimulants such as Vivarin® and No-Doz®, dieting aids such as Dexatrim® and Diatac®, and headache remedies such as Anacin®, Excedrin®, and Vanquish®.

Alkaloids are a broad category of nitrogen containing organic metabolites produced by plants. Since they contain nitrogen, they behave like bases (alkalis) and hence, they are termed alkaloids. Other common alkaloids are morphine, quinine, cocaine, and codeine. These substances are extremely bitter and/or toxic. By producing them, plants make their leaves unattractive to eating by insects and higher animals.

Like these substances, caffeine has a potent influence on our central nervous system. Hence, it is used as a stimulant in that morning cup of coffee or tea, or in a tablet of No-Doz® for a student studying chemistry late at night.

In this experiment, you will take a sample of coffee that has been brewed earlier by your instructor and extract the caffeine from it by using dichloromethane **(CH_2Cl_2)**. Dichloromethane is an organic solvent that has several useful properties for this extraction. It is insoluble in water, so it can be mixed with the coffee solution and then the two liquids may be easily separated. Additionally, the caffeine is much more soluble in it than in water. By adding dichloromethane to your coffee solution and mixing the two, the caffeine will dissolve out of the water and into the dichloromethane. The two liquids can then be separated and the dichloromethane, which has a very low boiling point, can be evaporated in the hood, leaving a residue of caffeine.

Experiment 18 The Extraction of Caffeine from Coffee

In order to verify that you have obtained caffeine, you will take a melting point of this residue. Pure caffeine melts at 238°C. The material that you have extracted will probably have a slightly lower melting point.

III. MATERIALS:

Coffee solution, 50 mL. 250 mL and 500 mL Erlenmeyer flasks, 100 mL graduated cylinder, sodium carbonate, dichloromethane, 250 mL beaker, 12.5 cm filter paper, iron ring, glass funnel, anhydrous sodium sulfate, 50 mL beaker, balance, pipet, hot plate, melting point apparatus, melting point tube, 4" test tube, label.

IV. PROCEDURE:

A. Extraction of the Coffee Solution:

1. To a clean 500 mL Erlenmeyer flask, add 100 mL of coffee extract. This extract will have been prepared for you in advance by the instructor.

2. Add approximately two grams of sodium carbonate **(Na_2CO_3)** to the coffee solution. This will react with some of the substances in the coffee extract to make them extremely water soluble and less likely to be extracted into the dichloromethane. Swirl the mixture until all the sodium carbonate dissolves.

3. Add 25 mL of dichloromethane **(CH_2Cl_2)**, and **vigorously** swirl the mixture for 10 minutes. **Do not shake the mixture or an emulsion will form.**

4. Allow the mixture to stand and separate into two layers- a dark aqueous top layer and a clear dichloromethane bottom layer.

5. Carefully pour as much of the top layer as you can into a beaker, without removing the bottom layer. This process is called decanting.

6. Place a 12.5 cm fluted filter paper in a long stem glass funnel. Put the funnel in a small iron ring and suspend it over a 250 mL Erlenmeyer flask.

7. Using a squeeze bottle of water, thoroughly wet the filter paper.

8. Slowly and carefully pour the dichloromethane/water mixture into the fluted filter paper. The excess water will drain through and the dichloromethane solution of caffeine will remain on the filter paper.

9. Using a pipet, transfer the dichloromethane solution to a 50 mL Erlenmeyer flask. To this solution, add a scoop of anhydrous sodium sulfate **(Na_2SO_4)** in order to remove the last traces of water.

10. While the solution is drying, weigh (tare) a 50 mL beaker to the nearest 0.001 g on a balance. Record this tare weight in **Table 1**.

11. Using a pipet, transfer the dried solution to the tared 50 mL beaker.

12. Evaporate <u>most</u> of the dichloromethane in the hood on a warm hot plate. When only a fraction of a milliliter of liquid is left, remove the beaker from the hot plate. Allow the beaker to stand in the hood for a minute or two. The heat remaining in the glass will cause the last amount of dichloromethane to evaporate and produce a solid residue of crude caffeine.

Experiment 18 The Extraction of Caffeine from Coffee

13. In order to determine your recovery of caffeine, reweigh the cool beaker and record this mass in the Table.

14. By difference, determine the mass of the caffeine in the beaker and record this value in the Table.

15. Pure caffeine is a white solid. Describe the appearance of your product in the Table.

16. Your instructor may ask you to check the purity of your caffeine by a melting point determination.

B. Melting Point Determination:

1. This description of the melting point determination involves the use of a Mel-Temp apparatus. If you are using a different apparatus, your instructor will make the appropriate changes in directions.

2. Add a small amount of your caffeine to a capillary melting point tube. This can most easily be done by pressing the open end of the tube down on some of the caffeine, turning the tube right-side-up, and tapping the tube on the bench until the solid falls to the bottom of the tube. Tap the tube on the bench a few more times so that the caffeine is compacted.

3. Place the tube in the slot of the melting point apparatus. Turn on the switch of the instrument. This will turn on a light that illuminates the sample. By looking through the magnifying glass, you should have a clear view of your sample.

4. Turn the heating dial to a setting of about 40. This will cause a rapid heating of your sample initially, but should not cause it to melt. As the temperature increases, there will be a decrease in the rate of heating. Since your sample will melt above 200°C, you may need to periodically increase the setting of the heating dial. Ideally, the temperature increase at the melting point should be only 1-2°C per minute.

5. Record the melting range of your sample as follows. In the Table record the temperature at which you first see liquid beginning to form. The sample will continue to melt. When the last of the sample has melted, record the second temperature. These two values are separated by a dash, i.e., 63-65°C. For a relatively pure compound, this temperature range will generally be only a few degrees.

6. If requested by your instructor, turn in your product in a labeled test tube.

Table 1

	Data
Tare weight of the empty beaker	g
Mass of the beaker and the caffeine	g
Mass of the caffeine	g
Appearance of the caffeine	g
Melting range of the caffeine	°C

Experiment 18 The Extraction of Caffeine from Coffee

Name:_____   Section:_____

Experiment 18 Problems

1. a. Is dichloromethane more or less dense than water?

   b. What evidence do you have to support your answer?

2. What is the melting point of pure caffeine? You may find this value in the *Handbook of Chemistry* or the *Handbook of Chemistry and Physics*.

3. Why do you think the melting point of your caffeine was different from the literature melting point?

4. Why was sodium carbonate added to the coffee solution?

5. Explain why the caffeine that you obtained in this experiment was not white in color.

6. List some substances that contain caffeine other than coffee.

   a.

   b.

   c.

   d.

   e.

Name:_____   Section:_____

EXPERIMENT 19

The Preparation of Soap

I. OBJECTIVES:

A. To understand how soap is prepared.

B. To prepare soap from a vegetable oil.

C To test some of the properties of soap.

II. DISCUSSION:

Over the ages, soap has played a dominant role as a cleaning agent. Its usefulness relates to the ability of soap molecules to interact with both polar water molecules and nonpolar greases and oils. By doing so, soap solutions in water provide substantially enhanced cleaning characteristics over that of water alone. Because soap molecules can interact with nonpolar oils, soap solutions can disperse tiny droplets of nonpolar oil in polar water to form an emulsion. Then, the water washes the emulsion away from the article being cleaned.

While the history of soap making may be extremely ancient, the Romans are generally credited with developing a procedure for soap making, involving the heating of goat fat with the extracts of wood ashes (bases). With the development of the modern chemical industry in the 1800's, right up to today, the general process has not changed fundamentally. Animal or vegetable fats or oils are heated with a base, such as sodium hydroxide, to produce soap and glycerol. This process is termed *saponification*.

$$\begin{matrix} CH_2OCR \\ | \\ CHOCR \\ | \\ CH_2OCR \end{matrix} \quad + \quad 3\ NaOH \quad \xrightarrow{\Delta} \quad 3\ NaOCR \quad + \quad \begin{matrix} CH_2OH \\ | \\ CHOH \\ | \\ CH_2OH \end{matrix}$$

a fat or oil   a fatty acid salt   glycerol
                                                  (soap)

The R groups in the fat or oil consist of long carbon chains with accompanying hydrogen atoms. An example is the saponification of a typical animal fat, tristearin.

Experiment 19 The Preparation of Soap

$$\begin{array}{c} CH_2OC(O)(CH_2)_{16}CH_3 \\ | \\ CHOC(O)(CH_2)_{16}CH_3 \\ | \\ CH_2OC(O)(CH_2)_{16}CH_3 \end{array} + 3\ NaOH \xrightarrow{\Delta} 3\ NaOC(O)(CH_2)_{16}CH_3 + \begin{array}{c} CH_2OH \\ | \\ CHOH \\ | \\ CH_2OH \end{array}$$

tristearin   sodium stearate   glycerol

Usually the fats and oils have different R groups in a molecule, and therefore give a mixture of different sodium salts upon saponification. In order to separate the soap from the rest of the reactants and products, the product mixture is treated with concentrated sodium chloride solution. This causes the soap to coagulate (salt out) and separate from the solution. It can then be collected by filtration and washed to purify it.

In this experiment you will treat olive oil with hot sodium hydroxide to produce a soap that consists of the sodium salts of the following fatty acids. The percentages indicate the contribution that they make to the composition of the olive oil. Thus, the soap that you make will consist mostly of sodium oleate.

$$CH_3(CH_2)_7\text{-CH=CH-}(CH_2)_7COOH \qquad \text{oleic acid (84 \%)}$$

$$CH_3(CH_2)_{14}COOH \qquad \text{palmitic acid (7 \%)}$$

$$CH_3(CH_2)_4\text{-CH=CH-}CH_2\text{-CH=CH-}(CH_2)_7COOH \qquad \text{linoleic acid (5 \%)}$$

$$CH_3(CH_2)_{16}COOH \qquad \text{stearic acid (2 \%)}$$

After you have prepared and collected your soap, you will test some of its properties and compare them to the properties of commercial soap. A measurement of pH will give you an idea of how effectively you were able to remove the sodium hydroxide. Both soaps will be used to emulsify an oil; an indication of how effective they are at removing dirt. Finally you will see the difference in adding soap to distilled water versus water containing a relatively high concentration of calcium ions ("hard" water). These ions, and also Mg^{+2} and Fe^{+3} ions, tend to react with soaps to form insoluble salts (the ring around the tub or the telltale gray on clothes). For example, a soap containing sodium stearate would give a precipitate of calcium stearate:

Experiment 19 The Preparation of Soap

$$2 \ CH_3(CH_2)_{16}\overset{\overset{O}{\|}}{C}ONa \ + \ Ca^{+2} \longrightarrow [CH_3(CH_2)_{16}\overset{\overset{O}{\|}}{C}O]_2Ca \ + \ 2 \ Na^{+1}$$

III. MATERIALS:

A 400 mL beaker, 100 mL graduated cylinder, olive oil, ethanol, 20% sodium hydroxide solution, glass rod, hot plate, saturated sodium chloride solution, 125 mL filtering flask, Filtervac, Buchner funnel, filter paper, 3 125 mL Erlenmeyer flasks, commercial soap, pH paper, mineral oil, 5% calcium chloride solution.

IV. PROCEDURE:

A. Preparation of the Soap:

1. To a 400 mL beaker, add 20 mL of olive oil, 20 mL of ethanol (**C₂H₅OH**), and 25 mL of 20 % sodium hydroxide (**NaOH**) solution. Stir the mixture with a glass rod.

2. Place the beaker on a hot plate. Caution: the alcohol is flammable.

3. Continue heating until you can no longer smell the odor of the alcohol. At this point, the reaction mixture should have turned into a pasty mass. Remove the beaker from the hot plate and allow the mixture to cool to room temperature.

4. When cool, add to the mixture about 100 mL of a saturated sodium chloride (**NaCl**) solution and stir the resulting mixture <u>thoroughly</u>. The soap should coagulate into a solid mass and can now be filtered to free it from the sodium hydroxide solution and the glycerol by-product.

5. Set up a filtering flask with a Buchner funnel, Filtervac, and filter paper. Turn on the source of vacuum and collect the soap. With the vacuum still applied, wash the soap with two 10 mL portions of ice cold water.

6. Transfer the soap to a paper towel and allow it to dry.

B. Testing the Soap:

1. Take a pea sized piece of your soap and place it in a 125 mL Erlenmeyer flask. Add about 50 mL of distilled water. To a second Erlenmeyer flask add a similar sized piece of commercial soap and 50 mL of distilled water. To a third flask add 50 mL of distilled water. Stopper the flasks, shake them for 20-30 seconds. Observe the results in terms of solubility of the soaps and the foaming action in each flask. Record your observations in **Table 1**.

2. Dip clean glass rods into each flask and touch them to pieces of pH paper. Record the approximate pH of each in the Table

3. To each of the three flasks add two drops of mineral oil. Stopper and shake the flasks for about 10 seconds. How effective are the solutions at emulsifying the oil? Record your results in the Table.

4. Clean all three flasks with soap and water. Rinse each with distilled water. Put a pea sized piece of your soap in one of the flasks and a piece of the commercial soap in the second flask. Add about 50 mL of distilled water to all three of the flasks. To each flask add 10 drops of a 5 % solution of calcium chloride

Experiment 19 The Preparation of Soap

(**CaCl₂**). Shake each flask for about 10 seconds and observe any changes that may occur. Record your observations in the Table.

Table 1

Test	Your Soap	Commercial Soap	Distilled Water
Solubility and Foaming			
pH			
Emulsification of mineral oil			
Response to CaCl₂			

Experiment 19 The Preparation of Soap

Name:_____   Section:_____

Experiment 19 Problems

1. What is an emulsion?

2. Explain what is meant by saponification.

3. Why was ethanol added to the reaction mixture?

4. a. In areas of the country where the water is "hard," what problem can occur when washing with soap?

   b. Explain the chemistry of this problem.

Experiment 19 The Preparation of Soap

5. Which was better at maintaining an emulsion?

                oil and water   or   oil, water, and soap

Explain why, in terms of molecular theory.

6. Explain how soap can remove oil from clothing.

7. a. How did the pH of your soap compare with that of commercial soap?

    b. How do you account for the difference?

8. Solutions of bases such as NaOH and KOH feel slippery when you get them on your fingers. Explain why.

Name:_____   Section:_____

EXPERIMENT 20
Radioactivity

I. OBJECTIVES:

A. To understand the process of radioactivity.

B. To measure radioactive emissions under different conditions.

C. To determine the half-life of a radioactive isotope.

II. DISCUSSION:

Near the end of the 19th century, it was found that the nuclei of the isotopes of some elements undergo a spontaneous decay to different nuclei through the release small particles and electromagnetic radiation. These isotopes are termed radioisotopes. This emission process is termed **radioactivity**. For example, iodine-131 undergoes a radioactive decay to yield xenon-131 and an electron.

$$^{131}_{53}I \rightarrow\ ^{131}_{54}Xe\ +\ ^{0}_{-1}e$$

In this **nuclear reaction,** the symbol for each particle is written to show its mass number as a superscript and its atomic number or nuclear charge as a subscript. Notice that the Law of Conservation of Mass applies to nuclear reactions, as does the conservation of electrical charge. The sum of charges on the right equals the charge on the left, i.e., the Atomic Number.

By 1903, three types of radioactive emission had been identified.

Alpha particle emission results in the release of a helium ion (an α particle) from a radioactive nucleus. For example, radium-222 undergoes α emission to produce a radon atom.

$$^{222}_{88}Ra\ \rightarrow\ ^{218}_{86}Rn\ +\ ^{4}_{2}He$$

Likewise, polonium-210 decays by the release of an α particle to produce a lead atom.

$$^{210}_{84}Po\ \rightarrow\ ^{206}_{82}Pb\ +\ ^{4}_{2}He$$

Notice that in both of these nuclear reactions, mass is conserved, i.e., the superscript on the left side of the equation equals the sum of the superscripts on the right side. Likewise, the subscripts equal each other.

Beta particle emission results in the release of a high energy electron (a β particle) from the nucleus. This can be viewed as a neutron being converted to a proton and an electron, and the latter then is ejected from the nucleus. For example, radium-228 undergoes spontaneous β particle emission to form an actinium atom.

$$^{228}_{88}Ra \rightarrow \, ^{228}_{89}Ac + \, ^{0}_{-1}e$$

Since a neutron is converted to an electron and a proton, the atomic number of the product has increased, while the mass number remains unchanged.

Gamma ray emission results in the release of a form of electromagnetic radiation (γ-rays) that are similar to X-rays, although higher in energy. Since only energy is released in this decay, the isotope retains its original identity; it simply changes from a higher energy nucleus to one possessing a lower energy.

The emission of these various types of radiation can be measured by the use of a Geiger counter. The general scheme of a Geiger counter is shown in **Figure 1**.

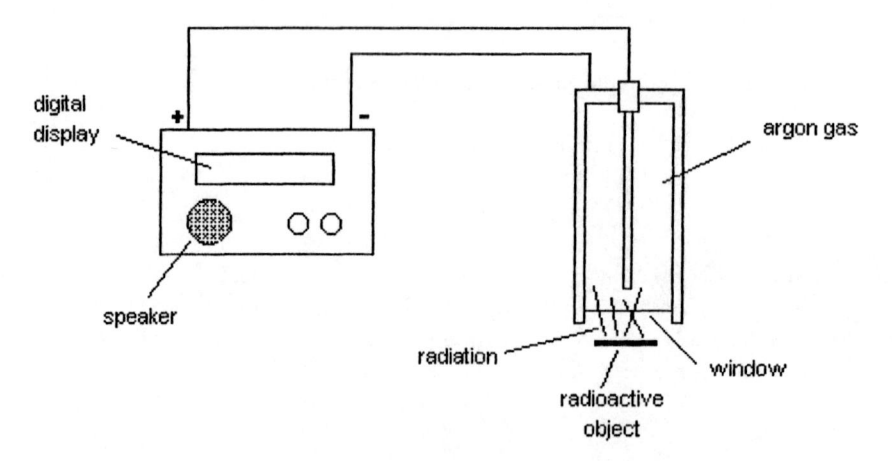

Figure 1

The radiation passes through the window of the probe and causes argon gas atoms to be ionized. This results in the flow of current to the positively charged electrode, causing a click to be heard from the speaker and an increase in the value shown in the LED digital display.

In this experiment, you will use a Geiger counter to measure the radioactive emissions from potassium compounds. Potassium consists of three naturally occurring isotopes. The isotope with mass number 40 is radioactive and emits β particles. By measuring potassium chloride at room temperature and at an elevated temperature, you will be able to determine the effect of temperature on the rate of radioactive decay. You will also measure the radioactive emissions of potassium sulfate and potassium phosphate to determine the effect of chemical combination on the rate of decay of potassium-40.

The rate of decay of radioisotopes varies considerably. Some radioisotopes decay rapidly in a few seconds, while others decay slowly over billions of years. In addition, not all of the atoms of a radioisotope decay at the same time. The time required for half the atoms of a radioisotope to decay is termed the **half-life**. For example, the half-life of iodine-131 is eight days. If you had a 100 g sample of pure I-131 today, the amount remaining after different periods of time is shown by the data in **Table 1**.

Experiment 20 Radioactivity

Table 1

Day	Amount of I-131 Remaining, g
0	100
8	50
16	25
24	12.5
32	6.2

After 10 half-lives, about 0.1 g of I-131 would remain. The rest would have decayed to Xe-131.

From your measurements of the radioactivity of potassium chloride, you will be able to estimate the half-life of potassium-40.

III. MATERIALS:

Geiger counter, two 100 mL beakers, potassium chloride, potassium sulfate, potassium phosphate, hot plate, thermometer, stopwatch.

IV. PROCEDURE:

A. The Determination of Background Radiation:

1. Obtain a Geiger counter, turn it on, and let it warm. Your instructor will demonstrate its operation and how to determine the reading of the radioactive emission from a sample.

2. Determine the level of background radiation in counts per minute by taking six one-minute measurements. Record this data in the **Table 2**.

B. The Determination of Radiation Emitted by Potassium Chloride at Room Temperature:

1. Label two 100 mL beakers as Beaker 1 and Beaker 2. Tare the beakers and record their tare weights.

    Beaker 1 Tare Weight: _____ g

    Beaker 2 Tare Weight: _____ g

2. Add enough solid potassium chloride (**KCl**) to each beaker to give a depth of solid of 1 cm. Reweigh the beakers and record their masses.

    Mass of Beaker 1 with KCl: _____ g

    Mass of Beaker 2 with KCl: _____ g

3. By difference, determine and record the masses of the potassium chloride in the two beakers.

    Mass of KCl Sample 1: _____ g

    Mass of KCl Sample 2: _____ g

4. Obtain a hot plate, turn it on to the lowest setting, and place the beaker containing KCl Sample 2 on it.

Experiment 20 Radioactivity

5. Determine the radiation count from the KCl Sample 1. Tap the beaker so that the surface of the KCl sample is level. Make sure that the Geiger counter probe is as close to the sample as possible, without actually touching the solid. As before, take six one-minute measurements and record them in the Table.

C. The Determination of Radiation Emitted by Potassium Sulfate at Room Temperature:

1. Clean out Beaker 1 and add enough solid potassium sulfate (K_2SO_4) to give a depth of solid of 1 cm. Weigh and record the mass of this beaker.

      Beaker 1 with K_2SO_4: _____ g

2. By difference, determine and record the mass of the potassium sulfate.

      Mass of K_2SO_4: _____ g

3. As before, make six one-minute measurements of the radiation emissions from the solid and record your data in the Table.

D. The Determination of Radiation Emitted by Potassium Phosphate at Room Temperature:

1. Clean out Beaker 1 again, and now add enough solid potassium phosphate (K_3PO_4) to give a depth of solid of 1 cm. Weigh and record the mass of this beaker.

      Beaker 1 with K_3PO_4: _____ g

2. By difference, determine and record the mass of the potassium phosphate.

      Mass of K_3PO_4: _____ g

3. Once again, make six one-minute measurements of the radiation emitted by this sample and record your data in the Table.

E. The Determination of Radiation Emitted by Potassium Chloride at Elevated Temperature:

1. By now the KCl Sample 2 should have warmed up. Place a thermometer in the solid and determine its approximate temperature.

      Temperature of KCl Sample 2: _____ °C

2. Again, take six one-minute measurements of radiation counts being emitted from the potassium chloride and record the data in the Table.

3. For each of the four sets of measurements, determine the average value and record these in the Table.

F. The Determination of the Net Activity of the Samples:

Subtract the average background count from each of the sample count averages to get a corrected sample count, i.e., the **Net Activity** of the samples. Record this in the Table.

Table 2

Trial	Radiation Emission, counts/minute							Net Activity
	1	2	3	4	5	6	Average	
Background								
KCl Sample 1								
KCl Sample 2								
K_2SO_4								
K_3PO_4								

Experiment 20 Radioactivity

Name:_____   Section:_____

Experiment 20 Problems

1. Convert the number of grams of the solids that you weighed out to moles.

   a. KCl Sample 1:

   b. KCl Sample 2:

   c. K_2SO_4:

   d. K_3PO_4:

2. Calculate the number of potassium atoms in each of the above samples. Remember that the gram formula weight of a compound contains Avogadro's number of formula units of that compound.

   a. KCl Sample 1:

   b. KCl Sample 2:

   c. K_2SO_4:

   d. K_3PO_4:

Experiment 20 Radioactivity

3. Since potassium consists of only 0.0118 % K-40, not all of the atoms in your samples were radioactive. Calculate the number of K-40 atoms in each of the above samples.

   a. KCl Sample 1:

   b. KCl Sample 2:

   c. K_2SO_4:

   d. K_3PO_4:

4. In the Table you determined the Net Activity for the four samples. In Question 3 you determined the number of K-40 atoms that resulted in this Net Activity. Determine the Net Activity that would have been expected for Avogadro's number of K-40 atoms for each of the samples.

   a. KCl Sample 1:

   b. KCl Sample 2:

   c K_2SO_4:

   d. K_3PO_4:

5. What is the effect of temperature on the emission of radiation from K-40?

6. What effect does the chemical combination of a radioactive element have on its rate of radioactive emission?

7. Use the data for KCl Sample 1 to determine the half-life of K-40. The half-life, $t_{1/2}$, can be determined from the following equation:

$$t_{1/2} = \frac{0.693\ N}{\text{rate}}$$

where N is the number of K-40 atoms in the KCl Sample 1 (determined in Question 3) and rate is the Activity of the sample multiplied by 5 (the factor 5 is used since the Geiger counter will only detect about 20 % of the radiation emitted).

8. Radiocarbon dating was used to prove that the Shroud of Turin was no older than 1200 AD. This evidence was based on measuring the ratio of C-14 to C-12 and C-13. The latter two isotopes are stable, while C-14 undergoes a natural radioactive decay. Write the nuclear equation for the emission of a beta particle from carbon-14.

9. If radium-226 undergoes alpha emission, what isotope is produced?

Experiment 20 Radioactivity

10. A radioisotope produces xenon-128 and a beta particle. What is this radioisotope? Write the complete nuclear reaction.

11. Complete the following nuclear reactions:

a.

$$^{213}_{83}Bi \rightarrow \,^{213}_{84}Po + \underline{}$$

b.

$$^{234}_{91}Pa \rightarrow \,^{234}_{92}At + \underline{}$$

c.

$$^{221}_{87}Fr \rightarrow \,^{217}_{85}At + \underline{}$$

12. a. Technetium-99 has a half-life of 6.0 hours. How many grams of a 5.00 g sample of Tc-99 remain after one day?

b. How long would it take this sample to disintegrate to 0.04 g?

Name:_____   Section:_____

EXPERIMENT 21

Analysis of Acids

I. OBJECTIVES:

A. Become familiar with volumetric methods of analysis.

B Perform chemical analysis by titration.

C. Determine the acetic acid content of vinegar.

II. DISCUSSION:

Most common substances are packaged with information about their chemical composition and often contain information about the actual chemical analysis of the substance. The consumer of bleach, vinegar, rubbing alcohol, cereal, bread, lunch meat, and milk can find information on the label about the amount of specific chemical compounds present in the substance. Often, the mass or percentage of that compound in the substance is provided. This information arises from quantitative chemical analysis by a number of techniques.

A common method of analysis involves using solutions of chemical reagent to react with the specific chemical compound in the sample being analyzed. If the concentration of the chemical reagent in the solution used for the analysis is known, as well as the volume needed to react with the specific compound of interest, the amount of a specific chemical compound in the sample can be determined. Since the volume and concentration of the reactive chemical are central to the analysis, the method is called volumetric analysis.

The amount of chemical substances is expressed chemically in moles. In volumetric analysis, molarity (a concentration unit defined as moles of solute per liter of solution) is a unit that relates the volume of a standard reagent to the number of moles of the reagent. The relationship is molarity (M) times volume in liters (V) equals number of moles (n):

$$M \times V = n$$

The acidity of vinegar is due to the acetic acid ($HC_2H_3O_2$) which is present. In the analysis performed here, sodium hydroxide solution is added to a 10 mL sample of vinegar until the acetic acid in the vinegar is completely reacted according to the equation below.

$$HC_2H_3O_2 + NaOH \rightarrow NaC_2H_3O_2 + H_2O$$

Experiment 20 Analysis of Acids

That neutralization point will be identified by using an indicator, phenolphthalein. At that point, the number of moles of acid is related to the number of moles of sodium hydroxide by the stoichiometry of the equation. In this case, there is a 1:1 mole ratio.

III. MATERIALS:

125 mL Erlenmeyer flask, 10 mL pipet, pipet bulb, buret, buret funnel, two 250 mL beakers, standardized sodium hydroxide solution, phenolphthalein indicator solution

IV. PROCEDURE:

1. Obtain about 150 mL of standardized sodium hydroxide (**NaOH**) from the stock bottle. Record the molarity of the sodium hydroxide in **Table 1**.

2. Obtain about 50 mL of vinegar. Record the brand name and the rated acidity of the vinegar in the table.

3. Set up the buret on a ring stand with a buret funnel for filling. Rinse the buret twice with about 10 mL of sodium hydroxide solution and then fill it above the zero mark. Remove the buret funnel. Place a waste container under the buret. Open the stopcock fully to force air bubbles out of the buret tip and to lower the liquid level below the zero mark.

4. Rinse a 10 mL pipet with vinegar and then transfer exactly 10.00 mL of vinegar to an Erlenmeyer flask that has been rinsed with distilled water. Record the volume in the Table. Add 3 drops of phenolphthalein to the vinegar.

5. For the first practice run, titrate rapidly to obtain an approximate volume of sodium hydroxide solution that will be needed. Record the initial liquid level in the buret in the **Table 2** in the Practice column. Titrate the vinegar by rapidly adding the sodium hydroxide solution until the phenolphthalein just turns pink and remains pink. Be sure to wash down the walls of the flask with distilled water as you approach the end point. Record the final liquid level in the Table. Calculate the volume of sodium hydroxide that was used to titrate the acetic acid in the vinegar.

6. Refill the buret with sodium hydroxide solution and record an initial buret reading. Discard the contents of the Erlenmeyer flask, rinse it several times with tap water and then with distilled water.

7. Pipet a 10.00 mL sample of vinegar into the flask. Add 3 drops of phenolphthalein and titrate the sample. For this trial, add the sodium hydroxide titrant quickly to within a few milliliters of the volume needed for the practice run. Then approach the end point slowly to obtain a very precise volume of sodium hydroxide solution for the titration.

8. Repeat the analyses until 3 trials in which the sodium hydroxide volumes are within 0.2 mL have been obtained.

8. Calculate the average volume of sodium hydroxide solution used for the three good trials and record this in Table 1. Using the molarity and the volume, calculate the number of moles of sodium hydroxide needed to titrate the acetic acid in the sample of vinegar.

9. Using the equation, calculate the number of moles of acetic acid in the sample of vinegar. From the number of moles of acetic acid and its formula weight, calculate the number of grams of acetic acid in the vinegar sample.

10. Assume that the density of the vinegar is 1.005 g/mL. Calculate the mass of vinegar solution in 10.00 mL.

11. Finally, calculate the mass percent acetic acid in vinegar.

$$\% \text{ Acetic Acid} = \frac{\text{Mass of Acetic Acid}}{\text{Mass of Vinegar}} \times 100\%$$

Table 1

Molarity of Sodium Hydroxide	
Brand of Vinegar	
Reported Acidity of Vinegar	
Volume of Vinegar Used	
Average Volume of NaOH Solution Used in Titrations	
Number of moles of NaOH in Average Volume	
Number of Moles of Acetic Acid in 10 mL of Vinegar	
Mass of Acetic Acid in 10 mL of Vinegar	
Mass of 10 mL of Vinegar	
% Acetic Acid in Vinegar	

Table 2

	Practice	Trial 1	Trial 2	Trial 3	Trial 4
Initial Buret Reading					
Final Buret Reading					
Volume of NaOH Added					

Experiment 20 Analysis of Acids

Name:_____   Section:_____

Experiment 21 Problems

1. Compare your experimental result for percent acetic acid in vinegar with the value listed on the label. Do the two values agree with each other? Account for differences in the values or differences in the way the values are reported.

2. Consider the following potential errors in the analysis of vinegar. Would the following factors cause the experimental percent acetic acid in vinegar to be too high or too low, or have no effect?

   a. The pipet used to measure the vinegar consistently retained too much vinegar.

   b. The sodium hydroxide solution was labeled with a molarity that was too low. Its actual concentration was higher.

   c. The buret retained a bubble of air in the tip and it remained a consistent size throughout the experiment.

   d. Extra water was added to the Erlenmeyer flask before the titraton with sodium hydroxide was started.

3. Oxalic acid, $H_2C_2O_4$, is used in various types of cleaners and in industrial applications. A 5.04 gram sample of a white solid is analyzed for oxalic acid content by titrating the oxalic acid with 0.267 M sodium hydroxide. The equation for the titration reaction is given below. The titration requires 26.24 mL of sodium hydroxide solution. What is the weight percent oxalic acid in the white solid?

$$H_2C_2O_4 \;+\; 2\,NaOH \;\rightarrow\; Na_2C_2O_4 \;+\; 2\,H_2O$$

4. An aspirin tablet contains a mixture of aspirin (acetylsalicylic acid, $HC_9H_7O_4$), starch, glycerol triacetate, and talc. The aspirin in two tablets is analyzed by titration of a solution of the two aspirin tablets with 0.1046 M potassium hydroxide, KOH, solution. The two aspirin tablets require 34.87 mL of the potassium hydroxide solution to completely react according to the following equation:

$$HC_9H_7O_4 + KOH \rightarrow KC_9H_7O_4 + H_2O$$

How many milligrams of aspirin are in each tablet?

5. Analysis of air samples is used to characterize the amount of pollutants that are present. To analyze for sulfur dioxide in a sample of air, a 50 liter sample of air was bubbled through a solution of hydrogen peroxide to convert the sulfur dioxide into sulfuric acid, H_2SO_4. The resulting sulfuric acid was titrated with 0.00426 M NaOH. The titration required 6.87 mL of NaOH to reach the end point.

$$H_2SO_4 + 2\,NaOH \rightarrow 2\,H_2O + Na_2SO_4$$

a) How many moles of NaOH were used in the titration?

b) How many moles of sulfuric acid were present in the sample of air?

c) How many moles of sulfur are present in the number of moles of sulfuric acid determined in b)?

d) How many grams of sulfur were present in the sample? In the 50 mL of air?

Name:_____   Section:_____

EXPERIMENT 22

Characteristics of Antacids

I. OBJECTIVES:

A. Learn the composition and characteristics of antacids.

B. Determine the capacity of antacids for neutralizing acids.

C. Characterize the rate of neutralization for antacids.

II. DISCUSSION:

Antacids are one of the largest over the counter drugs used by Americans. In the past, antacids were bases that neutralized excess stomach acid that was generated when an individual consumed foods that stressed the stomach. Nowadays antacids may be very sophisticated in their action. Some control the biological process by which acid is secreted into the stomach so that excess acid is prevented in the first place.

The stomach is a highly acidic environment where hydrochloric acid is secreted to aid in the digestion of food. Proteins and carbohydrates break down into their constituent amino acids and simple sugars, respectively, more rapidly in an acidic environment than in neutral solution. The acidity of the stomach has been characterized as being similar to 1 molar hydrochloric acid.

Antacids that act to neutralize excess stomach acid should have a number of characteristics. They should be fast-acting to provide rapid relief from the discomfort. An antacid that raises the pH of the stomach too high may lead to continued acid secretion and actually heighten the discomfort of acid indigestion. The antacid should also be present in sufficient quantity to neutralize the acid. Of course, the base in the antacid must not harm the mouth or esophagus during consumption.

Milk of magnesia, which is a slurry of magnesium hydroxide in water, will neutralize hydrochloric acid by the following reaction:

$$Mg(OH)_2 \; + \; 2\,HCl \; \rightarrow \; MgCl_2 \; + \; 2\,H_2O$$

One mole of magnesium hydroxide will neutralize two moles of hydrochloric acid. Because the magnesium hydroxide is not in solution, its reaction with the hydrochloric acid will be slower than when soluble sodium hydroxide reacts with hydrochloric acid. Sodium hydroxide would be too corrosive to use as an antacid.

This experiment consists of two parts. In Part A, the acid neutralizing capacity of a dose of antacid will be evaluated. Since the antacid may act slowly, the analysis will be performed as a **back titration**. A dose of antacid will be allowed to react with excess hydrochloric acid until the antacid is fully consumed. Excess

Experiment 22 Characteristics of Antacids

acid will remain. The amount of excess acid will be determined by titration with standard sodium hydroxide solution:

$$HCl + NaOH \rightarrow H_2O + NaCl$$

Since the total number of moles of acid added (AA) will equal the number of moles of acid that reacted with the antacid (AR) plus the number of moles of acid titrated (AT) with sodium hydroxide, the number of moles of acid that reacts with the antacid (AR) can be determined by subtracting the number of moles of acid which was titrated with sodium hydroxide (AT) from the total number of moles of acid added (AA).

Total Acid Added = Acid Reacted with Antacid + Acid Titrated with NaOH
    (AA)   =   (AR)   +   (AT)

or

Acid Reacted with Antacid = Total Acid Added - Acid Titrated with NaOH
    (AR)   =   (AA)   -   (AT)

In Part B of the experiment, the pH changes that occur when an antacid is added to 1 molar hydrochloric acid will be characterized. This is meant to simulate a situation where the antacid is added to an acidic stomach environment and then additional acid is secreted over a period of time. An effective antacid should rapidly react with the hydrochloric acid but not raise the pH too high. One reasonable way to rate the antacid is to characterize the length of time the solution remains in the range of pH 3-5.

II. MATERIALS:

Antacids, aluminum foil, buret, standardized hydrochloric acid solution (app. 0.5 M), standardized sodium hydroxide solution (app. 0.5 M), bromocresol green indicator solution, 2 150 mL beakers, 100 mL graduated cylinder, 250 mL Erlenmeyer flask, pH meter, stir bar, stirring motor

III. PROCEDURE:

A. Acid Neutralizing Capacity of Antacids

1. Obtain about 150 mL of standard hydrochloric acid **(HCl)** solution in a 250 mL beaker.

2. Choose one of the available antacids and record the brand, the identity of the active ingredient, and the antacid dose in **Table 1**. Measure out one antacid dose (the smallest dose specified on the label). If the antacid is a solid pill or pellet, fold it securely in several layers of aluminum foil and crush it with an iron ring.

3. Add the antacid to a 250 mL Erlenmeyer flask. Add 20 mL of water and 3 drops of bromocresol green indicator to the flask.

4. Use a pipet to measure 10.0 mL of hydrochloric acid and transfer it to the flask. Swirl the flask for s few minutes. Continue to add 10 mL aliquots of the hydrochloric acid, recording the number of 10.0 mL aliquots added, until there is distinct evidence that the antacid has completely reacted. (For carbonate-containing antacids, the evolution of carbon dioxide should cease and all clumps of solid should disperse. The audible fizzing upon swirling with excess acid is a good indicator of the continued reaction with the antacid. For hydroxide-containing antacids, the hydroxide in the reactant should completely dissolve and the solution should clarify significantly. The hydrochloric acid should react with all of the antacid and some hydrochloric acid should remain in the solution. The bromocresol green will turn yellow due to the excess acid, although this change is not a good indicator of when all of the antacid has reacted. Note the changes that occur during the reaction.)

Experiment 22 Characteristics of Antacids

5. Obtain 100 mL of standardized NaOH (**NaOH**) solution from the side shelf in a clean, dry beaker. Rinse a buret twice with 10 mL portions of the sodium hydroxide solution. Set up the buret on a stand and fill it above the zero mark. Place a waste container under the buret and open the stopcock fully to flush the air out of the tip and to bring the liquid level below the 0.00 mark.

6. Take a reading of the liquid level and record it in the Table. Titrate the excess acid until the bright yellow color disappears and a gray to green color remains for 15 seconds. Read the liquid level in the buret and record the value in the Table. (The end point is not as sharp as many acid-base titrations.)

7. From the volume of hydrochloric acid used and its molarity, calculate the number of moles of hydrochloric acid added to the antacid (AA).

8. From the volume and molarity of the sodium hydroxide used in the titration, calculate the number of moles of sodium hydroxide used in the titration. This is equal to the number of moles of hydrochloric acid titrated by the sodium hydroxide solution (AT).

9. By difference, calculate the number of moles of HCl neutralized by the antacid (AR).

10. Using its formula weight, convert moles of HCl to grams of HCl.

11. Repeat the titration for a second trial.

Table 1

	Trial 1	Trial 2
Brand of Antacid		
Active Ingredients		
Dose		
Molarity of Hydrochloric Acid		
Volume of Hydrochloric Acid		
Molarity of Sodium Hydroxide		
Initial Buret Reading		
Final Buret Reading		
Volume of Sodium Hydroxide Used in Titration		
Moles of Sodium Hydroxide Used in Titration		
Moles of Hydrochloric Acid Added (AA)		
Moles of Hydrochloric Acid Titrated By Sodium Hydroxide (AT)		
Moles of Hydrochloric Acid Reacted With Antacid (AR)		
Grams of Hydrochloric Acid Reacted With Antacid		

Experiment 22 Characteristics of Antacids

B. Characteristics of Neutralization by Antacids

1. Select an antacid and record the brand, the active ingredients, and the dose size in **Table 2**. If the antacid is a pill or tablet, crush it in an aluminum foil packet.

2. Place 100 mL of distilled water in a beaker. Add a stir bar. Set the beaker on a stirrer plate and adjust the stirrer so the solution stirs smoothly with the stir bar slightly off center. Immerse a pH electrode in the beaker away from the stir bar and set the pH meter to read the pH.

3. Record an initial pH reading in the Table (Volume HCl added is 0). Wait 30 seconds and record another pH reading.

4. Use a pipet to add 5.0 mL of standardized HCl solution. After 30 seconds, record a pH reading. Add another 5.0 mL aliquot of hydrochloric acid and take another pH reading after 30 seconds.

5. Add one dose of the antacid to the beaker and take a pH reading after 30 seconds.

6. Add 5.0 mL of the hydrochloric acid and record the pH reading after 30 seconds. Continue to add the 5.0 mL aliquots of hydrochloric acid and record pH readings until the pH reading is below 1.5. Add a minumum of 40 mL of acid.

7. When finished, rinse the pH electrode thoroughly and immerse it in the storage solution. Discard the reacton mixture and wash the glassware thoroughly and return to its proper location.

Table 2

Brand of Antacid		
Active Ingredients		
Dose		
	Volume of HCl added, mL	pH
Initial pH Reading (Before Antacid Added)	0	
pH After 30 Seconds	0	
pH after 1st aliquot of HCl added	5.0	
pH after 2nd aliquot of HCl added	10.0	
pH after antacid added	10.0	
pH after 5 more mL of HCl added	15.0	
etc.	20.0	
	25.0	
	30.0	
	35.0	
	40.0	
	45.0	
	50.0	
	55.0	
	60.0	

Experiment 22 Characteristics of Antacids

Name:_____   Section:_____

Experiment 12 Problems

1. List the antacids analyzed by the class in order of decreasing amounts of hydrochloric acid neutralized. List the active ingredients for each. Discuss the results.

Order of Acid-Consuming Capacity of Antacid	Ingredients
Greatest	
Least	

2. Alka Seltzer® contains sodium bicarbonate ($NaHCO_3$) as the active ingredient. When it reacts with an acid, carbon dioxide gas is evolved and salt and water are produced. Write a balanced equation for the reaction.

3. Maalox is an antacid which contains 400 mg of aluminum hydroxide, $Al(OH)_3$, and 200 mg of magnesium hydroxide, $Mg(OH)_2$, in each dose. What is the total number of grams of HCl would be neutralized by one dose of Maalox?

$$Al(OH)_3 + 3\,HCl \rightarrow 3\,H_2O + AlCl_3$$

$$Mg(OH)_2 + 2\,HCl \rightarrow 2\,H_2O + MgCl_2$$

Experiment 22 Characteristics of Antacids

3. Compare your data with that of a classmate who used a different antacid. On the next page, prepare a graph of pH vs volume of HCl added for the your data and that of your classmate.

4. Discuss the results as they relate to the speed of neutralization and the magnitude of the pH changes observed for each antacid tested.

a) Did either antacid raise the pH above 5?

b) Which antacid held the pH above 2 for the longest period of time?

c) Was there evidence to indicate when the antacid was completely consumed? Was there a significant drop in pH at that point?

Experiment 22 Characteristics of Antacids

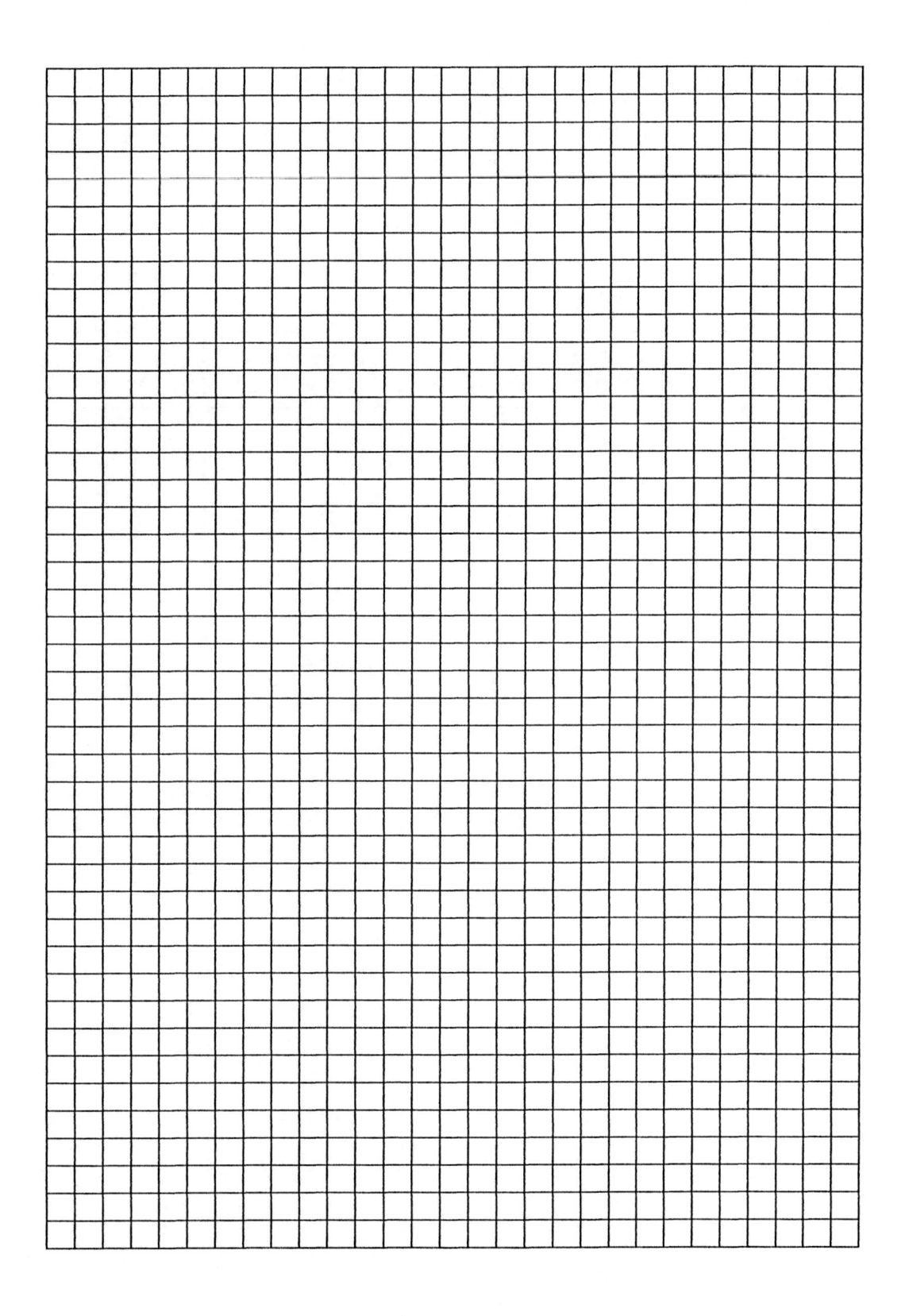

Appendix A
PERIODIC TABLE OF THE ELEMENTS

1 1A	2 2A	3 3B	4 4B	5 5B	6 6B	7 7B	8 8B	9 8B	10 8B	11 1B	12 2B	13 3A	14 4A	15 5A	16 6A	17 7A	18 8A
1 **H** 1.0079																	2 **He** 4.0026
3 **Li** 6.941	4 **Be** 9.0122											5 **B** 10.81	6 **C** 12.011	7 **N** 14.0067	8 **O** 15.9994	9 **F** 18.9984	10 **Ne** 20.179
11 **Na** 22.9898	12 **Mg** 24.305											13 **Al** 26.9815	14 **Si** 28.0855	15 **P** 30.9738	16 **S** 32.06	17 **Cl** 35.453	18 **Ar** 39.948
19 **K** 39.0983	20 **Ca** 40.078	21 **Sc** 44.9559	22 **Ti** 47.88	23 **V** 50.9415	24 **Cr** 51.996	25 **Mn** 54.9380	26 **Fe** 55.847	27 **Co** 58.9332	28 **Ni** 58.69	29 **Cu** 63.546	30 **Zn** 65.38	31 **Ga** 69.72	32 **Ge** 72.59	33 **As** 74.9216	34 **Se** 78.96	35 **Br** 79.904	36 **Kr** 83.80
37 **Rb** 85.4678	38 **Sr** 87.62	39 **Y** 88.9059	40 **Zr** 91.22	41 **Nb** 92.9064	42 **Mo** 95.94	43 **Tc** (98)	44 **Ru** 101.07	45 **Rh** 102.905	46 **Pd** 106.4	47 **Ag** 107.868	48 **Cd** 112.41	49 **In** 114.82	50 **Sn** 118.69	51 **Sb** 121.75	52 **Te** 127.60	53 **I** 126.904	54 **Xe** 131.29
55 **Cs** 132.905	56 **Ba** 137.33	57* **La** 138.905	72 **Hf** 178.49	73 **Ta** 108.948	74 **W** 183.85	75 **Re** 186.207	76 **Os** 190.2	77 **Ir** 192.22	78 **Pt** 195.08	79 **Au** 196.966	80 **Hg** 200.59	81 **Tl** 204.383	82 **Pb** 207.2	83 **Bi** 208.980	84 **Po** (209)	85 **At** (210)	86 **Rn** (222)
87 **Fr** (223)	88 **Ra** 226.025	89 **Ac**# 227.028	104 **Rf** (261)	105 **Db** (262)	106 **Sg** (263)	107 **Bh** (262)	108 **Hs** (265)	109 **Mt** (268)	110 **Uun** (269)	111 **Uum** (272)	112 **Uub** (277)		114 **Uug** (277)		116 **Uuh**		

Lanthanides

58 **Ce** 140.115	59 **Pr** 140.907	60 **Nd** 144.24	61 **Pm** (145)	62 **Sm** 150.36	63 **Eu** 151.965	64 **Gd** 157.25	65 **Tb** 158.925	66 **Dy** 162.50	67 **Ho** 164.930	68 **Er** 167.26	69 **Tm** 168.934	70 **Yb** 173.04	71 **Lu** 174.967

Actinides

90 **Th** 232.038	91 **Pa** 231.036	92 **U** 238.029	93 **Np** 237.048	94 **Pu** (244)	95 **Am** (243)	96 **Cm** (247)	97 **Bk** (247)	98 **Cf** (251)	99 **Es** (252)	100 **Fm** (257)	101 **Md** (258)	102 **No** (259)	103 **Lr** (260)

Appendix B

Conversion Factors

A. Metric Prefixes

giga	G	$1{,}000{,}000{,}000 = 10^9$
mega	M	$1{,}000{,}000 = 10^6$
kilo	k	$1000 = 10^3$
deci	d	$0.1 = 10^{-1}$
centi	c	$0.01 = 10^{-2}$
milli	m	$0.001 = 10^{-3}$
micro	μ	$0.000001 = 10^{-6}$
nano	n	$0.000000001 = 10^{-9}$

B. English-Metric Conversions

1 mile	=	1.6093 km
1 inch	=	2.54 cm
1 pound	=	453.6 g
1 quart	=	0.946 L